Engineering
for Mathematicians

James Nutaro

AMERICAN ACADEMIC PRESS

AMERICAN ACADEMIC PRESS

By AMERICAN ACADEMIC PRESS

201 Main Street

Salt Lake City

UT 84111 USA

Email manu@AcademicPress.us

Visit us at http://www.AcademicPress.us

ISBN: 979-8-3370-8954-6

Distributed to the trade by National Book Network Suite 200, 4501 Forbes Boulevard, Lanham, MD 20706

10 9 8 7 6 5 4 3 2 1

My aim in writing this book is to introduce models of engineered systems to persons with a background in mathematics but little or no experience in engineering or physics. The idea emerged from a graduate course that I taught in the mathematics department at the University of Memphis. Using a method similar to what I present here, my students finished the semester being able to look at an engineering diagram and write its governing equations.

If you work with engineers or in topic areas related to engineering, then it is my hope that this book will be useful to you. Models in engineering are living things. They evolve as our understanding of a problem grows, as new features are added to a design, and as new requirements or experiences expose a system's limitations. The more actively you can engage in this process of evolution, the more opportunities you will have to apply your knowledge and skills toward a project's success. This is good for you, good for your team, and good for the users of the system you build.

The method of modeling that I present is, in its essence, the method of bond graphs. There are many excellent books on bond graphs written for engineers. An advantage when teaching bond graphs to engineers is that the method can be learned by analogy. The engineering student comes prepared with an understanding of how to build models in their domain of study. From this starting point, it is enough to expose the underlying principle of conservation of power; the interchangeability of elements like springs and capacitors, dampers and resistors, and such and then use examples to drive the point home.

This book differs from others on bond graphs in that I do not assume a background in engineering. Instead, we begin with the axiomatic scheme embedded in the method of bond graphs, without concern for its relation to physical machines. Once this scheme is firmly grasped, we proceed by stages toward its physical interpretation. With practice, it is my hope that you will develop the intuitive insights that an engineer enjoys when constructing a model.

If you study other texts on bond graphs, you will see that I have altered some aspects of the subject and omitted others. I have simplified, or perhaps only changed, the graphical notation. The standard treatment builds on familiar (to an engineer) symbology from the engineering domains. In this book, a system of notation is built from scratch with the goal of making it more intelligible to readers without a background in engineering.

I have omitted the graphical analysis of causality and causal marks. The purpose of causal analysis is (in my view) to aid in simulation. Causal analysis distinguishes, for instance, a graph that produces ordinary differential equations from a graph that produces differential algebraic equations. Of course, this distinction is apparent in the model's equations. I suspect that for students who are new to the method and possess a high degree of mathematical maturity, the graphical analysis of causality engenders more confusion than clarity.

If you find the material in this book intriguing, useful, or (maybe!) both, then I encourage you to seek out one or more of the many excellent books on bond graphs written for engineers. With the knowledge you have gained here, I hope that you will find these books to be approachable and informative and that they will expand your reach when applying the method.

I wish to express my gratitude to Dr. Vladimir Protopopescu for his insightful criticism of my draft manuscripts, the time we spent discussing revisions, and for his twenty years of mentorship at Oak Ridge National Laboratory.

Contents

1. Fundamentals

Our modeling method uses a directed graph to describe the components of a system and their relations to one another. Engineers call the edges of this graph bonds or power bonds. The graph is called a bond graph. The physical motivation for this name won't interest us for several chapters yet. Therefore, I will refer to our graphs simply as graphs and use the familiar terms edge and node (or vertex) to describe its parts.

Edges connect the components of a system. Two variables are associated with each edge: the effort e and flow f. Our graphical notation for an edge is the labeled arrow

$$\xrightarrow{1}$$

The effort for this edge is e_1 and the flow is f_1.

The vertices of the graph model components and points of connection. Equations governing the efforts and flows are imposed by the vertices attached to each edge. The resulting system of equations models the behavior of the system.

0 and 1 junctions A zero junction has degree two or more. The edges attached to a zero junction have equal efforts. The sum of flows pointing into the junction equals the sum of flows pointing out of the junction. For example, the graph

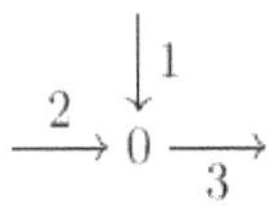

assigns the relations

$$e_1 = e_2 = e_3$$
$$f_3 = f_1 + f_2$$

If we change the edge orientations such that

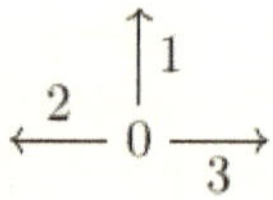

then the relations become

$$e_1 = e_2 = e_3$$
$$f_1 + f_2 + f_3 = 0$$

To offer one more example, the graph

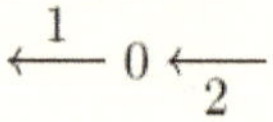

produces the relations

$$e_1 = e_2$$
$$f_1 = f_2$$

A one junction is the same as a zero junction but with the role of flow and effort reversed. The edges attached to a one junction have equal flows and the sum of the incoming efforts equals the sum of the outgoing efforts. For example, the graph

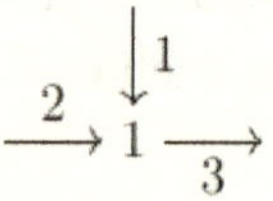

assigns the relations

$$f_1 = f_2 = f_3$$
$$e_3 = e_1 + e_2$$

Basic elements A surprising number of engineered devices can be modeled with five elements: SE, SF, E, F, and Z. These elements are vertices with degree one. A vertex that defines $e(t)$ as a function of time is a source of effort. These vertices are labeled SE. A source of flow defines $f(t)$ as a function of time. These vertices are labeled SF. The direction of the arrow matters! Our graphs will always have the edge pointing away from sources of effort and sources of flow.

The elements E, F, and Z relate effort and flow on their adjacent edge. Each has a parameter E, F, or Z according to the type of element. The direction of the arrow matters! Our graphs will always have the edge pointing into E, F, and Z elements. The relations imposed by these elements are

$$e(t) = e(0) + \frac{1}{E}\int_0^t f(\tau)\, d\tau \text{ , by an E element}$$

$$f(t) = f(0) + \frac{1}{F}\int_0^t e(\tau)\, d\tau \text{ , by an F element}$$

$$e(t) = Zf(t)\text{, by a Z element}$$

We label the vertex with E, F, or Z respectively to indicate the relation that is imposed.

For the E and F vertices, we prefer, when possible, to have $e(t)$ and $f(t)$ in their differential form[1]

$$\dot{e}(t) = f(t)/E \text{ , } e(0) = e_0$$
$$\dot{f}(t) = e(t)/F \text{ , } f(0) = f_0$$

[1] We use dot notation throughout this text, wherein $\dot{x} = dx/dt$, $\ddot{x} = d^2x/dt^2$, and so forth.

If we can write our equations in this way then our model will be a system of differential equations, possibly with algebraic constraints.

Systems of equations Given a graph we can extract a system of equations that corresponds to the relations imposed by its vertices. A complete graph produces a number of equations equal to the number of unknown variables. Physical systems can be mapped to complete graphs, and a model that fails to produce a complete graph is almost certainly in error.

Several examples will illustrate the process of writing equations from a complete graph. The graph

$$\mathrm{E} \xleftarrow{1} 0 \xleftarrow[2]{} \mathrm{SF}$$

has for its governing equations

$$\begin{aligned} e_1 &= e_2 \\ f_1 &= f_2 \\ \dot{e}_1 &= f_1/E \end{aligned}$$

This can be reduced to the single equations $\dot{e}_1 = f_2/E$ where f_2 is defined by the source of flow.

An oscillator can be defined with an E, F, and SE element such that

$$\begin{array}{ccc} \mathrm{E} \xleftarrow{1} & 1 & \xrightarrow[2]{} \mathrm{F} \\ & {\scriptstyle 3}\uparrow & \\ & \mathrm{SE} & \end{array}$$

The system of equations for this graph is

$$
\begin{aligned}
f_1 &= f_2 = f_3 \\
e_1 + e_2 &= e_3 \\
\dot{e}_1 &= f_1/E \\
\dot{f}_2 &= e_2/F
\end{aligned}
$$

We can define $f = f_1 = f_2 = f_3$ to obtain the pair of equations

$$
\begin{aligned}
\dot{e}_1 &= f/E \\
\dot{f} &= (e_3 - e_1)/F
\end{aligned}
$$

where e_3 is supplied by the SE element.

An uncommon type of system is defined with an E and SE element such that

$$
\mathrm{SE} \xrightarrow{1} 1 \xrightarrow[2]{} \mathrm{E}
$$

Taking $e = e_1 = e_2$ and $f = f_1 = f_2$ we obtain the governing equation $\dot{e} = f/E$. Because e is prescribed by the source of effort, our unknown is the flow f. In principle, we have sufficient information to solve the problem. In practice, the appearance of these types of relations often, but not always, indicates a modeling error.

The graph below incorporates each of the elements discussed so far.

$$
\begin{array}{ccccccc}
 & & \mathrm{F} & & \mathrm{E} & & \\
 & & 2\uparrow & & \uparrow 5 & & \\
\mathrm{SE} & \xrightarrow[1]{} & 1 & \xrightarrow{4} & 0 & \xleftarrow[6]{} & \mathrm{SF} \\
 & & \downarrow 3 & & & & \\
 & & \mathrm{Z} & & & &
\end{array}
$$

Using the relationships defined by each element, we may write a system of equations for this graph.

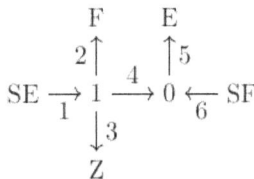

$$
\begin{aligned}
&e_2 + e_3 + e_4 = e_1 && \text{the 1 junction} \\
&f_1 = f_2 = f_3 = f_4 && \\
&\dot{f}_2 = e_2/F && \text{the F element} \\
&e_3 = Zf_3 && \text{the Z element} \\
&f_5 = f_4 + f_6 && \text{the 0 junction} \\
&e_4 = e_5 = e_6 && \\
&\dot{e}_5 = f_5/E && \text{the E element}
\end{aligned}
$$

Many of the equations in this system are redundant. By rearranging and simplifying, this system is reducible to the pair of ordinary differential equations

$$
\begin{aligned}
\dot{e}_5 &= (f_2 + f_6)/E \\
\dot{f}_2 &= (e_1 - e_5 - Zf_2)/F
\end{aligned}
$$

The source of effort supplies e_1 and the source of flow f_6.

Often there will be several SF, SE, F, Z, and E elements in the graph. These can be distinguished by subscripts. For example, the graph

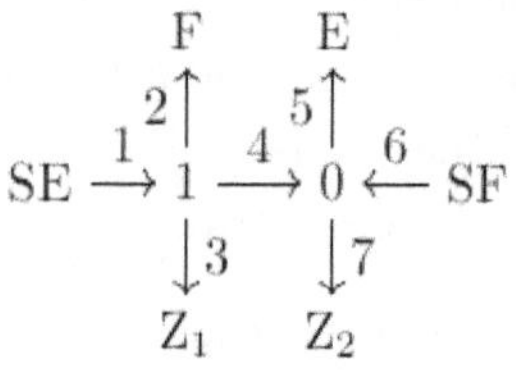

The system of equations for this graph is

$$
\begin{array}{ll}
e_1 = e_2 + e_3 + e_4 & \text{the 1 junction} \\
f_1 = f_2 = f_3 = f_4 & \\
\dot{f}_2 = e_2/F & \text{the F element} \\
e_3 = Z_1 f_3 & \text{the } Z_1 \text{ element} \\
f_4 + f_6 = f_5 + f_7 & \text{the 0 junction} \\
e_4 = e_5 = e_6 = e_7 & \\
e_7 = Z_2 f_7 & \text{the } Z_2 \text{ element} \\
\dot{e}_5 = f_5/E & \text{the E element}
\end{array}
$$

Rearranging and simplifying reduces this system to the pair of ordinary differential equations

$$
\dot{e}_5 = (f_2 + f_6 - e_5/Z_2)/E
$$
$$
\dot{f}_2 = (e_1 - e_5 - Z_1 f_2)/F
$$

The flow f_6 is supplied by the SF element and the effort e_1 by the SE element.

Transformers and gyrators The final two types of vertices we will consider in this chapter are the transformer T and gyrator G. These vertices have degree two. The arrangement of edges adjacent to the transformer is

$$
\xrightarrow{1} \mathrm{T} \xrightarrow{2}
$$

The transformer always has an edge pointing into the element and an edge pointing out of the element.

If the transformer is linear then it is characterized by a constant parameter T. The equations imposed by a transformer are

$$
T f_1 = f_2
$$
$$
T e_2 = e_1
$$

In these equations, the subscript 1 always indicates the inward pointing and 2 the outward pointing edge. Careful attention to units is necessary when defining the constant of transformation T. We will concern ourselves with that problem when a physical interpretation is introduced.

The gyrator resembles the transformer in having two edges such that

$$\xrightarrow{1} \mathrm{G} \xrightarrow{2}$$

It relates efforts and flows by

$$\begin{aligned} e_2 &= Gf_1 \\ e_1 &= Gf_2 \end{aligned}$$

when the gyrator is linear. Defining the constant G requires careful attention to units, and we will turn to this problem when a physical interpretation is introduced.

To illustrate the use of these elements, consider the graph

$$\mathrm{E}_1 \xleftarrow{1} 1 \xrightarrow{2} \mathrm{G} \xrightarrow{3} 1 \xrightarrow{4} \mathrm{E}_2$$

The system of equations for this graph is

$$\begin{aligned} f_1 &= f_2 \\ e_1 + e_2 &= 0 \\ f_3 &= f_4 \\ e_3 &= e_4 \\ \dot{e}_1 &= f_1/E_1 \\ \dot{e}_4 &= f_4/E_2 \\ e_3 &= Gf_2 \\ e_2 &= Gf_3 \end{aligned}$$

This system can be simplified to obtain

$$\dot{e}_1 = \frac{1}{GE_1} e_4$$
$$\dot{e}_4 = -\frac{1}{GE_2} e_1$$

Summary This small collection of elements will allow us to model a surprising diversity of physical systems. Of course, much of the physical world is not so simple as to be reduced to these forms. Nonetheless, many types of engineering diagrams depict systems that can be expressed in this way.

The graph elements and their rules originate in physical laws of conservation. However, we do not need to be concerned immediately with the origins of these rules to make use of the graphs. Instead, we can treat the graph properties as axioms that are true in each of the engineering domains that we will consider. These axioms are:

1. The sum of efforts into a one junction equals the sum of efforts out of a one junction.
2. The flows adjacent to a one junction are equal.
3. The sum of flows into a zero junction equals the sum of flows out of a zero junction.
4. The efforts adjacent to a zero junction are equal.
5. E, F, Z, SE, and SF elements are adjacent to exactly one edge.
6. The edge points into E, F, and Z elements.
7. The edge points away from SE and SF elements.
8. G and T elements have exactly two adjacent edges with one edge pointing in and one edge pointing out.

9. Each element imposes its equations on the variables of its adjacent edges.

Necessarily, these axioms restrict our study to graphs that satisfy them. For example, the graph

$$\mathrm{SE} \xrightarrow{1} \mathrm{SF}$$

is outside the scope of our study. The SF element is adjacent to an edge that violates axiom 7.

Exercises Write the equations in a reduced form for each of the graphs. Begin by producing the system of equations. Then simplify and rearrange to eliminate redundant terms.

Ex. 1.1

$$\begin{array}{ccccc} & & Z & & \\ & & \uparrow{\scriptstyle 3} & & \\ E & \xleftarrow{1} & 1 & \xrightarrow[2]{} & F \end{array}$$

Ex. 1.2

$$\begin{array}{ccccc} & & Z & & \\ & & \uparrow{\scriptstyle 3} & & \\ E & \xleftarrow{1} & 0 & \xrightarrow[2]{} & F \end{array}$$

Ex. 1.3

$$\begin{array}{ccccccccc} & & Z_1 & & Z_2 & & Z_3 & & \\ & & \uparrow{\scriptstyle 3} & & {\scriptstyle 5}\uparrow & & {\scriptstyle 8}\uparrow & & \\ E & \xleftarrow{1} & 1 & \xrightarrow[4]{} & 0 & \xrightarrow[6]{} & 1 & \xrightarrow{7} & F \\ & & {\scriptstyle 2}\uparrow & & & & & & \\ & & \mathrm{SF} & & & & & & \end{array}$$

Ex. 1.4

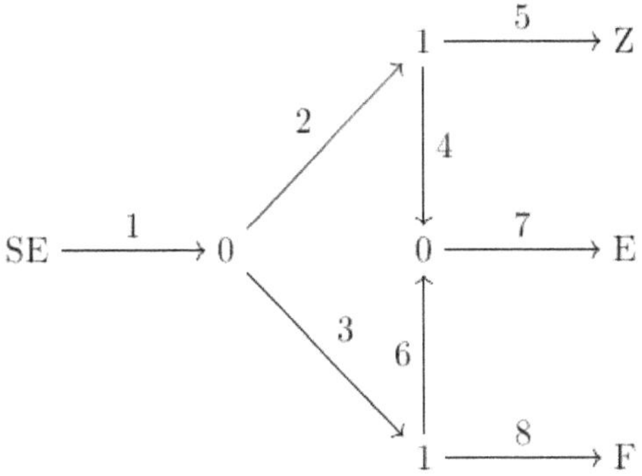

Ex. 1.5

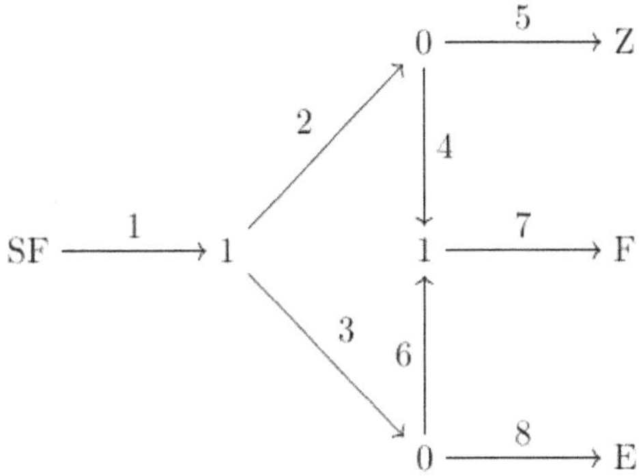

Ex. 1.6

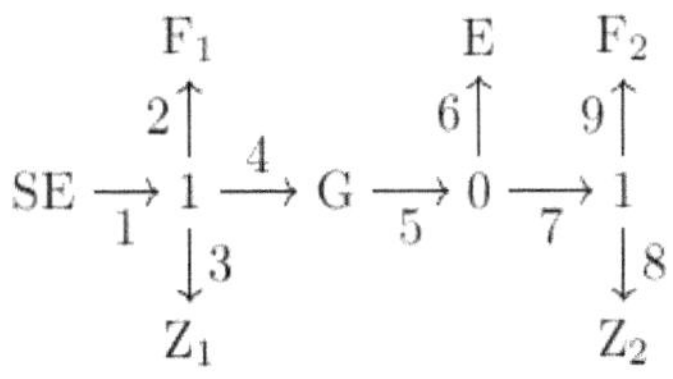

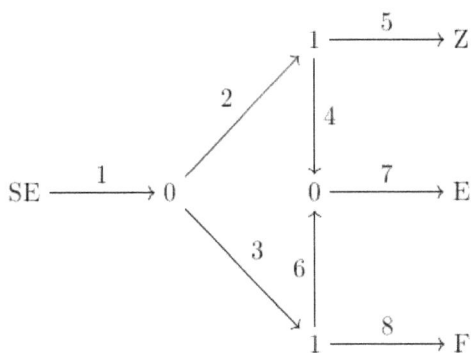

Ex. 1.7

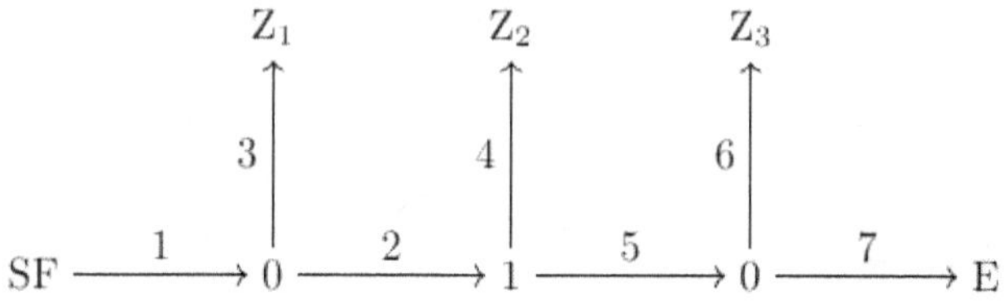

Ex. 1.8

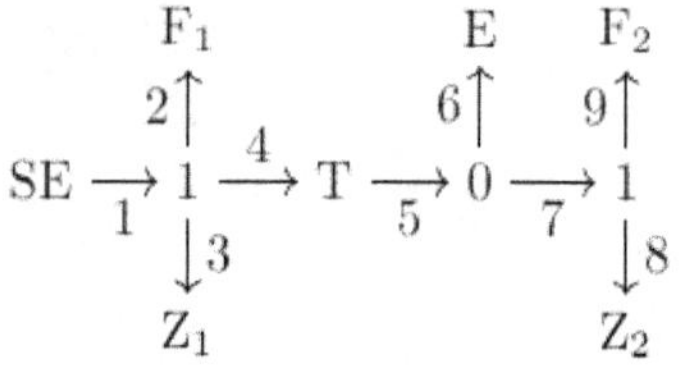

Ex. 1.9

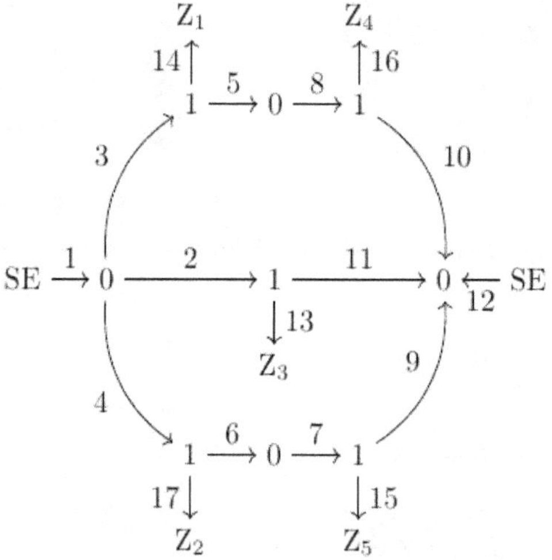

2. Other properties

Graphs have several properties that arise naturally from physical considerations, but which we can derive from mathematical principles without requiring (yet!) any physical insights.

p and q It will be useful to introduce two new variables p and q such that

$$p(t) = p(0) + \int_0^t e(\tau)\, d\tau$$

$$q(t) = q(0) + \int_0^t f(\tau)\, d\tau$$

In derivative form

$$\dot{p}(t) = e(t)\,,\, p(0) = p_0$$
$$\dot{q}(t) = f(t)\,,\, q(0) = q_0$$

Until we have a physical motivation for these variables, we will call p the accumulated effort and q the accumulated flow.

Memristor With accumulated effort and accumulated flow we can introduce the memristor. The memristor is indicated by a vertex M with degree one. The edge always points into the memristor. This element relates p and q according to the nonlinear relation

$$\dot{p}(t) = M(q(t))\dot{q}(t)$$

Because $\dot{p}(t) = e(t)$ and $\dot{q}(t) = f(t)$ we see that the relation is

$$e(t) = M(q(t))f(t)$$

This is a Z like element that changes the magnitude of its effect with the accumulated flow. In practice, the function M is nonlinear, but over a very small range of q the value of M will be approximately constant so that

$$e(t) = Mq(t)f(t)$$

The name memristor (short for memory resistor) comes from the field of electronics, but the element has analogs in other domains.

Other nonlinear elements The role of a vertex is to impose a relationship between the variables e, f, p, and q of its adjacent edges. The vertices we have defined so far will take us a long way, but there will be circumstances where others are needed. For instance, one way valves appear in many domains. The role of the one way valve is to prevent a negative flow resulting from a negative effort.

A model with two parameters I and α in which

$$f = I(\exp(\alpha e) - 1)$$

impedes flow when $e < 0$. The sharpness of the transition between impeding and admitting flow is controlled by α. The parameter I modulates the amplitude of the flow. This is the Shockley diode equation, which is used to model semiconductor diodes in electric circuits when the flow is not saturated; that is, before the flow has reached the limit of the device.

In other physical domains, we find other models of a valve. For instance, in hydraulics we can find a Z like vertex where

$$f = \begin{cases} \alpha K\sqrt{e} & e \geq 0 \\ -\alpha K\sqrt{-e} & e < 0 \end{cases}$$

The parameter $\alpha \in [0,1]$ describes how open the valve is (0 fully shut, 1 fully open) and K depends on physical properties of the fluid.

A nonlinear relation will appear in the system of equations produced from the graph. These relations may offer significant challenges for simulation and analysis, but they do not alter our modeling method.

Equivalent graphs The system of equations produced from a graph can be simplified by rearranging the graph to obtain smaller, equivalent structures. Several common cases are examined here. Consider

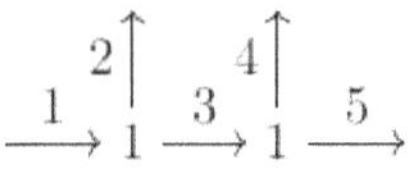

The flows and efforts in this graph are

$$\begin{aligned} f_1 &= f_2 = f_3 = f_4 = f_5 \\ e_1 &= e_2 + e_3 \\ e_3 &= e_4 + e_5 \end{aligned}$$

The terms e_3 and f_3 are defined by other quantities and are redundant. The above set of equations can be rewritten without e_3 and f_3 as

$$\begin{aligned} f_1 &= f_2 = f_4 = f_5 \\ e_1 &= e_2 + e_4 + e_5 \end{aligned}$$

The graph for these simplified equations is

$$\begin{array}{c}
2\uparrow \\
\xrightarrow{\;1\;} 1 \xrightarrow{\;5\;} \\
\downarrow 4
\end{array}$$

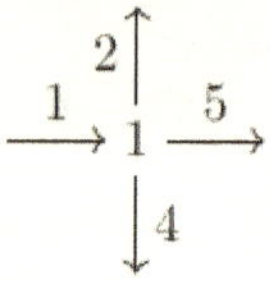

Given a graph fragment comprised entirely of one junctions, repeated application of this procedure will remove all but a single junction. A similar result can be derived for connected zero junctions.

Series components A set of components connected to a common one junction are said to be connected in series. Consider the graph

$$\begin{array}{c}
Z_1 \\
2\uparrow \\
\xrightarrow{\;1\;} 1 \xrightarrow{\;4\;} \\
\downarrow 3 \\
Z_2
\end{array}$$

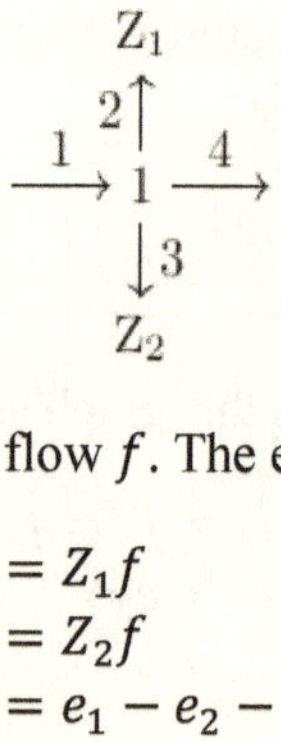

This graph has a single flow f. The efforts are

$$e_2 = Z_1 f$$
$$e_3 = Z_2 f$$
$$e_4 = e_1 - e_2 - e_3$$

Rearranging, we can write

$$e_4 = e_1 - (Z_1 + Z_2)f$$

We may define $Z = Z_1 + Z_2$ and the equivalent graph is

$$\begin{array}{c}
Z \\
\uparrow \\
\xrightarrow{\;1\;} 1 \xrightarrow{\;4\;}
\end{array}$$

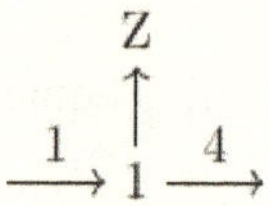

Now consider two E elements connected to the same one junction.

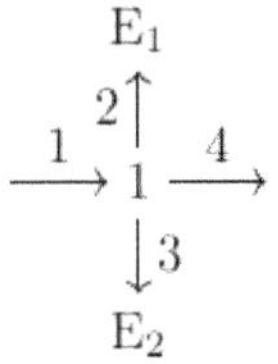

Again, we have a single flow f and the system of equations is

$$\begin{aligned}\dot{e}_2 &= f/E_1\\ \dot{e}_3 &= f/E_2\\ \dot{e}_4 &= \dot{e}_1 - \dot{e}_2 - \dot{e}_3\end{aligned}$$

We simplify to find

$$\dot{e}_4 = \dot{e}_1 - f(1/E_1 + 1/E_2)$$

Define

$$\frac{1}{E} = \frac{1}{E_1} + \frac{1}{E_2}$$

and so

$$E = \frac{E_1 E_2}{E_1 + E_2}$$

The equation written with E is

$$\dot{e}_4 = \dot{e}_1 - f/E$$

and the equivalent graph is

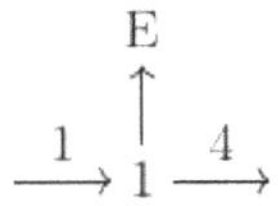

The problem of two F elements connected to a single one junction is left as an exercise for the reader.

Parallel components A set of components connected to a common zero junction are said to be connected in parallel. Consider the graph

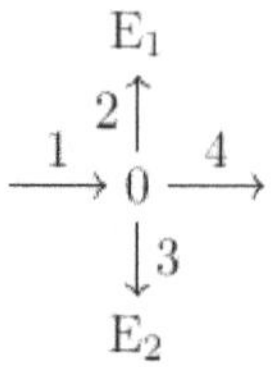

This graph has a single effort e. Its system of equations is

$$E_1\dot{e} = f_2$$
$$E_2\dot{e} = f_3$$
$$f_1 = f_2 + f_3 + f_4$$

A rearrangement yields

$$f_1 = \dot{e}(E_1 + E_2) + f_4$$

which is

$$\dot{e} = (f_1 - f_4)/(E_1 + E_2)$$

Define $E = E_1 + E_2$ and the equivalent graph is

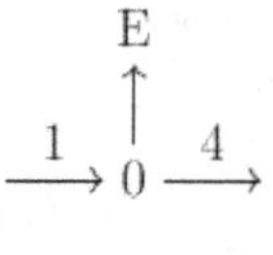

Symmetry suggests that two elements Z_1 and Z_2 attached to the same zero junction should produce an equivalent element

$$\frac{1}{Z} = \frac{1}{Z_1} + \frac{1}{Z_2}$$

and likewise for two elements F_1 and F_2.

Other equivalent structures A structure commonly appearing in practice is the diamond

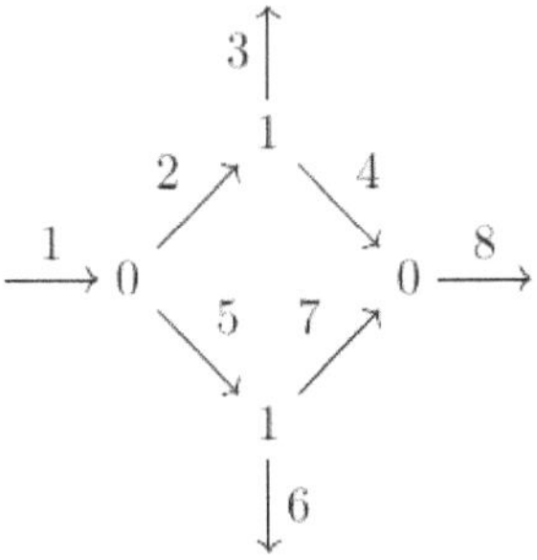

The system of flow equations for this graph is

$$
\begin{aligned}
f_1 &= f_2 + f_5 \\
f_2 &= f_3 = f_4 \\
f_5 &= f_6 = f_7 \\
f_4 + f_7 &= f_8
\end{aligned}
$$

Selecting f_3 and f_6 as our variables leaves us with

$$
\begin{aligned}
f_1 &= f_3 + f_6 \\
f_3 + f_6 &= f_8
\end{aligned}
$$

Now consider the effort equations. We have

$$
\begin{aligned}
e_1 &= e_2 = e_5 \\
e_2 &= e_3 + e_4 \\
e_5 &= e_6 + e_7 \\
e_7 &= e_8 = e_4
\end{aligned}
$$

We select e_1 and e_8 to obtain

$$
\begin{aligned}
e_1 &= e_3 + e_8 \\
e_1 &= e_6 + e_8
\end{aligned}
$$

Collecting our reduced systems of equations we have

$$f_1 = f_3 + f_6$$
$$f_3 + f_6 = f_8$$
$$e_1 = e_3 + e_8$$
$$e_1 = e_6 + e_8$$

A graph that produces this system is

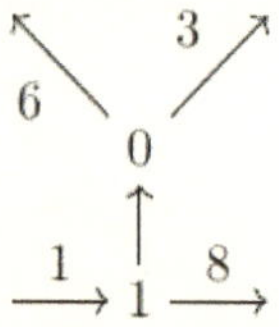

We have found two equivalent graph structures. A similar simplification, in which the one and zero junctions are interchanged, is left as an exercise for the reader.

Exercises

Ex. 2.1 Show that

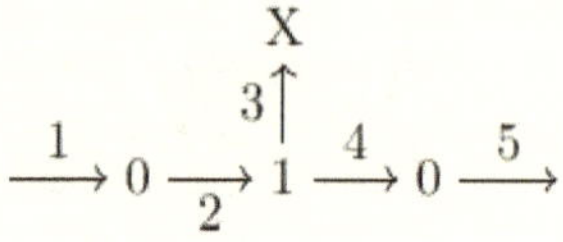

is equivalent to

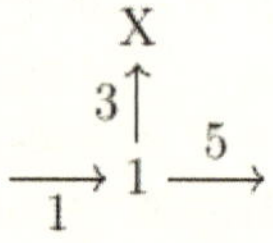

where X is any element with degree one.

Ex. 2.2 Show that

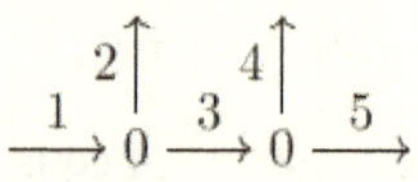

is equivalent to

$$\begin{array}{ccc} & \uparrow{\scriptstyle 2} & \\ \xrightarrow{1} & 0 & \xrightarrow{5} \\ & \downarrow{\scriptstyle 4} & \end{array}$$

Ex. 2.3 A rectifier is used to transform a sinusoidal input into a more or less constant output. Let the element D be a Schockley diode with parameters I and α. Let SE be an effort $e(t) = A\cos(t)$ with A a constant. Write equations for the graph

$$\begin{array}{ccccccc} & & \mathrm{D} & & \mathrm{E} & & \\ & & \uparrow{\scriptstyle 2} & & \uparrow{\scriptstyle 5} & & \\ \mathrm{SE} & \xrightarrow{1} & 1 & \xrightarrow{3} & 0 & \xrightarrow{4} & \mathrm{Z} \end{array}$$

Sketch the effort e_5 after several cycles of the source of effort have passed (hint: after the E element has stored some energy). What is the mechanism that sustains a positive value for e_5?

Ex. 2.4 A filter removes sediment from fluid flowing through a pipe. Many filters do this by trapping particles of sediment in a membrane that has small pores. Write equations for two variants of the graph below with S a source of flow in one variant and source of effort in the other. Assume M has a constant parameter M.

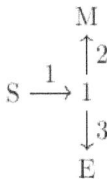

$$\begin{array}{ccc} & & \mathrm{M} \\ & & \uparrow{\scriptstyle 2} \\ \mathrm{S} & \xrightarrow{1} & 1 \\ & & \downarrow{\scriptstyle 3} \\ & & \mathrm{E} \end{array}$$

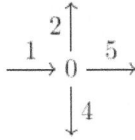

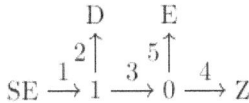

Sketch the effort across the M element as a function of time for each variant. Assume $q_2(0) > 0$. What happens to the M element when subjected to a source of flow? What happens when there is a source of effort?

Ex. 2.5 Show that a graph fragment consisting solely of zero junctions can be reduced to an equivalent graph with a single zero junction.

Ex. 2.6 Show that two elements F_1 and F_2 attached to the same one junction are equivalent to a single F element with parameter $F = F_1 + F_2$.

Ex. 2.7 Show that two elements Z_1 and Z_2 attached to the same zero junction are equivalent to a single Z element with parameter $1/Z = 1/Z_1 + 1/Z_2$.

Ex. 2.8 Show that two elements F_1 and F_2 attached to the same zero junction are equivalent to a single F element with parameter $1/F = 1/F_1 + 1/F_2$.

Ex. 2.9 In our previous example of the diamond shaped graph, derive a similar reduction after replacing each 1 with 0 and each 0 with 1. Hint: efforts and flows change places in the system of equations.

Ex. 2.10 Show that the two sources of effort in the graph below can be replaced with a single effort $e = e_1 + e_2$. Replace the one junction with a zero junction and the sources of effort with sources of flow. Show that the two sources of flow can be replaced with a single flow $f = f_1 + f_2$.

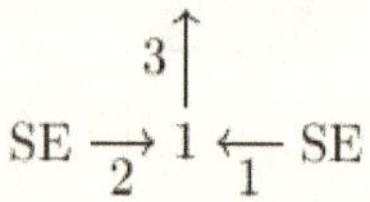

Ex. 2.11 Reduce the graph in exercise 1.9 to an equivalent graph with a single Z element.

3. Power and energy

Power and energy are fundamental physical quantities. In every domain, the effort and flow will be such that the power P flowing through an edge is the product

$$P(t) = e(t)f(t)$$

The energy Q accumulated at an element is

$$Q(t) = Q(0) + \int_0^t P\,(t)\,dt$$

It follows that power is the rate of change of energy

$$\dot{Q}(t) = P(t)\,,\ Q(0) = Q_0$$

Sources The source elements SE and SF produce and absorb power. Suppose a source of effort is a positive constant. Then the source produces power when $f > 0$ and absorbs power when $f < 0$. Likewise, a source of constant positive flow produces power when $e > 0$ and absorbs power when $e < 0$.

In some domains, and particularly when discussing electrical power systems, it can be convenient to introduce a new nonlinear element SP, a source of power. This element has degree one with its single edge pointing outward. The effort and flow on the outgoing edge satisfy

$$P = ef$$

Power dissipation The equation for power at a Z element is

$$ef = Zf^2 = e^2/Z$$

The squared term is positive. Therefore, the power flowing into the Z element is positive. The energy accumulated at the element cannot be recovered. This energy is lost to the system or, to be more precise, it has changed into a form that is not relevant to our model.

Most often, the lost energy becomes heat that dissipates into the environment. However, it is also possible that we have omitted some other system that consumes energy but is not relevant to the engineering problem at hand.

Power storage The equation for power at an E element is

$$P(t) = f(t)e(t) = f(t)\left(e(0) + \frac{1}{E}\int_{0}^{t} f(\tau)\, d\tau \right)$$

Consider a region where $e(t) \geq 0$. In this region, if $f(t) > 0$ then power is accumulated as the integral grows. Likewise, if $f(t) < 0$ then power is discharged as the integral diminishes. Should $e(t)$ become negative, then the E element accumulates power when $f(t) < 0$ and discharges power when $f(t) > 0$.

The F element also stores or discharges power in this way with the roles of e and f reversed. The power at an F element is

$$P(t) = e(t)f(t) = e(t)\left(f(0) + \frac{1}{F}\int_{0}^{t} e(\tau)\, d\tau \right)$$

If $f(t) > 0$ then power is accumulated when $e(t) > 0$ and discharged when $e(t) < 0$.

Power can also be expressed in terms of accumulated effort p and accumulated flow q. Replacing the integrals in each expression we find for the E element that

$$P(t) = f(t)(e(0) + q(t)/E - q(0))$$

and for the F element

$$P(t) = e(t)(f(0) + p(t)/F - p(0))$$

If power is positive, then the device is accumulating energy. Similarly, if power is negative, then the device is supplying energy. Consequently, E and F elements model devices that store energy.

Power passing through Zero junctions, one junctions, transformers, and gyrators transmit power. To see this, we start with the zero junction. A zero junction has a single effort e. Without loss of generality, we may assume the edges point into the junction. Multiplying the flow equation by the single effort we obtain the power equation

$$ef_1 + ef_2 + \cdots + ef_n = 0$$

The power entering the junction sums to zero just as the flows do.

The gyrator has power $e_1 f_1$ flowing into it and power $e_2 f_2$ flowing out. The equations imposed by the gyrator require that

$$\begin{aligned} e_2 f_2 &= G f_1 f_2 \\ e_1 f_1 &= G f_2 f_1 \end{aligned}$$

It follows from the second equation that

$$f_1 f_2 = e_1 f_1 / G$$

and so from the first equation

$$e_2 f_2 = G(e_1 f_1 / G) = e_1 f_1$$

Therefore, the power flowing into the gyrator equals the power flowing out of the gyrator. These derivations for the one junction and transformer are left as an exercise for the reader.

Units of measurement The International System of Units, known by the abbreviation SI units, is the most common system of measurement.[2] Time is measured in seconds (s). Distance is measured in meters (m), with one meter equal to 3.28 feet. Velocity is meters traveled per second and has units of m/s. Acceleration is change of velocity per second and has units of (m/s)/s = $\mathrm{m/s^2}$.

Mass (not weight!) is measured in kilograms (kg). Mass is a measure of the quantity of material (or, to be more precise, the inertia of the material); weight is a measure of force. The distinction is important! Force is measured in Newtons (N).

---

[2] And the most rational devised so far. If you have the misfortune to work with English or Imperial units, you will discover such quantities as slugs (for mass), two distinct types of pound (one for force, one for mass), two distinct types of ounce (again, one for force and another for mass), cups and pints and gallons (measures of volume), knots (a measure of velocity), and other colorfully named units related to each other by historical accident. To convert between these varied units, you must memorize, lookup, or calculate the appropriate multiplier each time you need it. You will make mistakes. SI units within a category (force or velocity, for example) are related by multiples of 10; for example, one kilometer is equal to 1,000 meters. This greatly simplifies mixing units to get, for example, power or volume flow. Use SI units.

One Newton is the amount of force needed to accelerate one kilogram at a rate of one meter per second squared. Therefore, $\mathrm{N} = \mathrm{kg} \cdot \mathrm{m/s}^2$. One Newton is approximately 0.225 pounds force.

Energy is measured in Joules (J). One Joule is the amount of energy expended when one Newton of force is applied to displace one kilogram by one meter. Hence, we have J = N·m. Energy is the precise expression of work done, hence the familiar refrain that work is force times distance. Power is measured in Watts (W) and, as energy per unit time, has units of J/s.

Exercises

Ex. 3.1 Show that the power entering a one junction equals the power leaving the one junction.

Ex. 3.2 Show that the power entering a transformer equals the power leaving the transformer.

Ex. 3.3 Let accumulated flow q have units of distance. Show that the flow f has units of velocity and that $\dot{f}$ has units of acceleration.

Ex. 3.4 Create (possibly fictional) units for e and f such that the product ef has units of power.

Ex. 3.5 Let $f(0) = 0$. Show that the energy stored in an F element is

$$Q = \frac{1}{2} F f^2$$

Hint: use $e = F\dot{f}$ and integrate by parts.

Ex. 3.6 Using the above expression for energy, show that units of mass for F and velocity for f are consistent with Joules for Q. What are the units of e so that ef is power?

Ex. 3.7 Let $e(0) = 0$. Show that the energy stored in an E element is

$$Q = \frac{1}{2}Ee^2$$

Ex. 3.8 Show that the total energy stored in the graph

$$\mathrm{E} \longleftarrow 1 \longrightarrow \mathrm{F}$$

is constant.

Ex. 3.9 Show that the total energy stored in the graph

$$\begin{array}{c} \mathrm{Z} \\ \uparrow \\ \mathrm{E} \longleftarrow 1 \longrightarrow \mathrm{F} \end{array}$$

is strictly diminishing for $Z > 0$. What happens if $Z < 0$?

Ex. 3.10 Consider the graph below.

$$\mathrm{SE} \longrightarrow 1 \longrightarrow \mathrm{F}$$

Let the initial flow $f(0) = 0$, the initial accumulated flow $q(0) = 0$, and e be a constant. Show that the energy stored in the F element when calculated as a function of the accumulated flow q is eq.

Ex. 3.11 A mass in the gravitational field of a planet (or other large mass) experiences a constant pull toward the planet's center. If the mass is lifted above the planet's surface and then falls it will be accelerated downward toward the surface at a rate g. Let our F element have units of mass. Show that the energy stored in the F element displaced upward by a distance h above the surface is Fgh. Hint: h is the accumulated flow, $h = 0$ is the surface of the planet, and the time derivative of f equals g; that is, $\dot{f} = g$.

Ex. 3.12 In the previous exercise, we release the mass and it falls toward the surface. Show that the sum of *potential* energy Fgh and *kinetic* energy $Ff^2/2$ is constant during the fall. Hint: $\dot{h} = f$. Use $f(0) = 0$.

4. Hydraulics

Hydraulics is the behavior of incompressible fluids that fill a confined space, such as a pipe. The dictionary's definition is "the branch of science and technology concerned with the conveyance of liquids through pipes and channels, especially as a source of mechanical force or control."[3] Hydraulic systems are common in construction equipment, robots, and other machines requiring precise and powerful motive forces. Hydraulics also play an important role in plumbing.

Flow f in a hydraulic system is the volume of fluid moving past a point per unit time, which we call volume flow. Volume flow has units of m^3/s. The accumulated flow q is volume in cubic meters. Effort e is pressure. Pressure has units of force per unit area, which is $\mathrm{N/m}^2$. One $\mathrm{N/m}^2$ is called a Pascal. Multiplying effort and flow we get

$$ef = \frac{\mathrm{N}}{\mathrm{m}^2} \cdot \frac{\mathrm{m}^3}{s} = \frac{\mathrm{N} \cdot \mathrm{m}}{s} = \frac{\mathrm{J}}{\mathrm{s}} = \mathrm{W}$$

0 and 1 junctions The zero junction is a joint that connects sections of pipe, hose, or other channels. Because the joint does not store or leak fluid, we expect that the flow into the joint equals the flow out of the joint. Hence, the sum of the flows in equals the sum of flows out. There is a single pressure at the joint.

A one junction is a section of pipe, hose, or channel. There is a single flow through the pipe. If e_1 is the sum of pressures

[3] From Oxford Languages online dictionary, accessed February 2025.

at the entrance to the pipe and e_2 the sum of pressures at the exit from the pipe then conservation of power requires

$$fe_1 = fe_2$$

Hence, the sum of the efforts at the inlet must equal the sum of the pressures at the outlet.

Basic elements The Z element is a resistance to flow. Physical elements that resist flow are valves, shear forces induced by the walls of a pipe or hose, and a membrane such as a filter. From its governing equation we know that the units of the Z element are

$$Z = \frac{e}{f} = \frac{\text{pressure}}{\text{volume flow}}$$

The E element stores fluid under pressure. Examples are reservoirs of every kind, for example: tanks, bladders, and columns of fluid. Using its governing equation, we find that the units of the E element are

$$E = \frac{f}{\dot{e}} = \frac{\mathrm{m}^3}{\mathrm{s}} \cdot \frac{\mathrm{s} \cdot \mathrm{m}^2}{\mathrm{N}} = \frac{\mathrm{m}^3 \cdot \mathrm{m}^2}{\mathrm{N}} = \frac{\text{volume}}{\text{pressure}}$$

An example of an E element is a quantity of fluid stored in a cylindrical column. The column has a cross-sectional area A, height h, and volume $V = Ah$. If the fluid has a density of ρ kilograms per unit volume, then the mass of fluid is ρV kg. Checking the units we see that

$$\rho V = \rho A h = \frac{\mathrm{kg}}{\mathrm{m}^3} \cdot \mathrm{m}^2 \cdot \mathrm{m} = \mathrm{kg}$$

Gravity acts on this column to create pressure at its bottom. Acceleration due to gravity g on Earth is 9.8 $\mathrm{m/s^2}$.

Multiplying g and ρV gives us units of kg·m/s^2, which is a Newton! Therefore

$$\frac{g\rho V}{A} = \frac{\text{N}}{\text{m}^2} = \text{Pascals}$$

is the pressure at the bottom of the column.

Plugging this expression for pressure into the expression for units of E gives us

$$E = \frac{\text{volume}}{\text{pressure}} = \text{m}^3 \cdot \frac{A}{g\rho V} = \frac{A}{g\rho}$$

where the m^3 in the numerator cancels the volume V in the denominator. We have derived the parameter E when the physical element is a column of fluid!

The governing equation of the F element expresses Newton's law of force, which states that the acceleration of a mass depends on the force applied to it. In hydraulics, this mass is the fluid in our pipe and the force is exerted by pressure acting on the surface of this incompressible mass of fluid.

We can see Newton's law of force in the units of F. From the governing equation we have

$$F = \frac{e}{\dot{f}} = \frac{\text{N}}{\text{m}^2} \cdot \frac{\text{s}^2}{\text{m}^3} = \frac{\text{N}}{\text{m}^4} \cdot \frac{1}{\text{acceleration}}$$

and then rearranging we get

$$\text{N} = (F \cdot \text{m}^4) \cdot \text{acceleration}$$

This expression contains force, acceleration, and a mass like term $F \cdot \text{m}^4$. Showing that the mass like term has units of kilograms is left for the reader.

A source of effort models pressure applied to the system that comes from outside the system. A typical source of pressure is the pressure exerted by the atmosphere on the open end of a section of pipe. For example, consider the faucet of a sink. When water flows out of the faucet, it is flowing into the air.

If you are standing near sea level, then the air around you exerts a pressure of 1 atmosphere, which is equal to 101,325 Pascals. This pressure is caused by the mass of the air above you being pulled downward by gravity, thereby applying a force to every surface exposed to the air. In our example of the faucet, the surface being pressed upon is the surface of the water flowing out of the faucet.

A source of flow models fluid flowing into the system at a rate unrelated to the pressures within that system. A hose placed in a bucket is an example of a source of flow. So is a pump that pushes water into the system. When modeling flow as a source, it must be reasonable to approximate the rate of flow as being unperturbed by changes in pressure.

Drawings to graphs The difficult task is to translate a drawing of a system into its graph. Having done so, we can extract a system of equations and simplify as before. The general procedure is as follows. First, assign a direction of flow through each element in the drawing and through each section of pipe or hose. Fluid flows from areas of high pressure to areas of low pressure, and so that is a natural choice for the direction of flow.

It is helpful to be consistent. Reversing the direction of flow midway along a series of pipe sections creates needless equalities that must be eliminated by rote algebra. Such choices do not invalidate our set of equations, but they complicate the problem of reducing the system of equations to a manageable form.

Having selected the directions of flow, create a graph that reflects the shape of the diagram. Each element in the diagram - pump, drain, tank, bladder, valve, and so forth - becomes a one junction with the corresponding E, F, Z, SF, or SE element attached.

Each section of tube, hose, and channel that connects elements becomes a zero junction between the one junctions of those elements. A zero junction is also introduced where several pipes are joined. If we have a choice of direction for an edge going into or out of a junction (1 or 0), then the edge should point in the direction of flow that we have chosen. Several examples will illustrate the procedure.

Leaky bucket Figure 1 illustrates a leaky bucket. Fluid is poured into the bucket, which is labeled E. The top of the bucket is open to the atmosphere and so the pressure at the surface of the fluid is one atmosphere (assuming the bucket is near sea level). As fluid accumulates, the pressure at the bottom of the bucket increases. The fluid under pressure then flows through a constricted pipe that offers resistance Z to flow. The flow emerges into the open air to exit at atmospheric pressure.

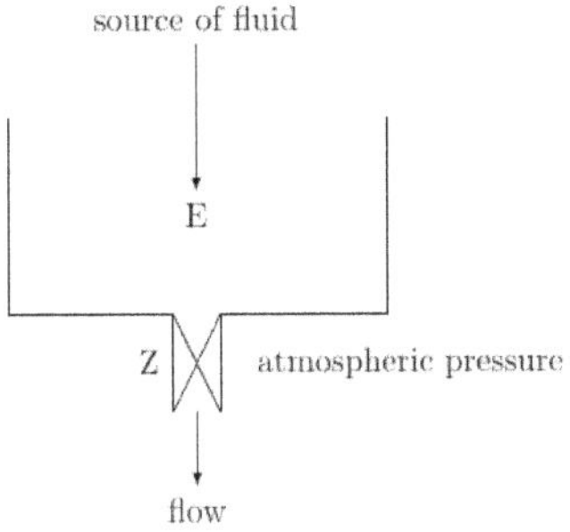

Figure 1. A leaky bucket.

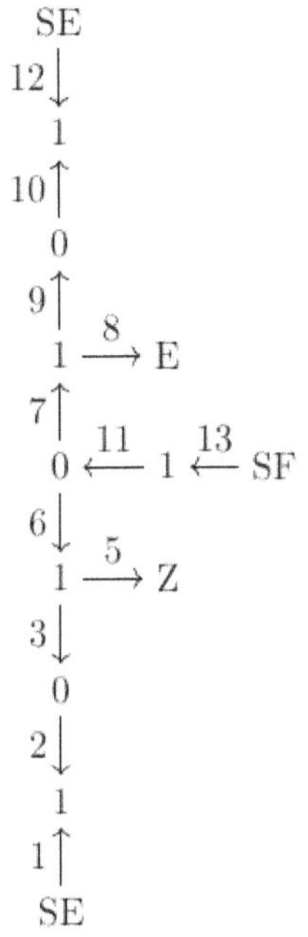

The graph for this diagram is shown above. Each one junction corresponds to a single element in the drawing: the atmospheric pressure SE, the fluid flowing into the bucket SF, the bucket capacity E, and the constricted pipe Z. At each point where elements in the drawing are connected there is a zero junction.

The source of flow is connected to the *bottom* of the E element! Fluid accumulates in the bucket from the bottom up. This connection to the bottom is literal if we imagine placing a hose in the bucket such that it rests on the bucket's floor. In this case, it is clear that fluid accumulates at the bottom as new fluid forces the existing fluid upward. However, the situation is no different if we place the hose at the top or middle of the bucket.

The position of the source of flow has implications for the direction of flow. The fluid flows downward through the pipe. Hence, the edge from the central zero junction to the one junction of the Z element points downward. However, the fluid flows upward into the bucket. Hence, the edge towards the E element points up.

We complete the graph by applying the axioms of direction. The edges attached to the E and Z nodes point into the elements. The edges attached to the sources point away from those elements. Where we have a choice, the edge points in the direction consistent with our assumed orientation of the fluid's flow: up into the bucket and down through the pipe.

The graph contains several structures that can be simplified at a glance. To begin, we see that

$$
\begin{aligned}
e_{10} &= e_9 \\
f_{10} &= f_9 \\
e_3 &= e_2 \\
f_3 &= f_2 \\
e_{11} &= e_{13} \\
f_{11} &= f_{13}
\end{aligned}
$$

We can discard edges 2, 10, and 11 to obtain the graph

$$
\begin{array}{rl}
 & \mathrm{SE} \\
12 & \big\downarrow \\
 & 1 \\
9 & \big\uparrow \\
 & 1 \xrightarrow{8} \mathrm{E} \\
7 & \big\uparrow \\
 & 0 \xleftarrow{13} \mathrm{SF} \\
6 & \big\downarrow \\
 & 1 \xrightarrow{5} \mathrm{Z} \\
3 & \big\downarrow \\
 & 1 \\
1 & \big\uparrow \\
 & \mathrm{SE}
\end{array}
$$

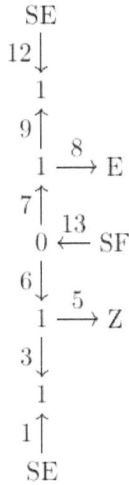

Consider the edges 1 and 3. We have

$$
\begin{aligned}
f_1 &= f_3 \\
e_3 + e_1 &= 0
\end{aligned}
$$

We are free to choose the direction of the 3 edge, and our choice was an inconvenient one. Reversing the direction of 3 gives us the simple relation $e_3 = e_1$. The same argument can be made for edges 9 and 12. Having reversed the directions of 3 and 9, it is immediately apparent that these are redundant. Removing them gives us the graph

$$
\begin{array}{rl}
 & \mathrm{SE} \\
12 & \big\downarrow \\
 & 1 \xrightarrow{8} \mathrm{E} \\
7 & \big\uparrow \\
 & 0 \xleftarrow{11} \mathrm{SF} \\
6 & \big\downarrow \\
 & 1 \xrightarrow{5} \mathrm{Z} \\
1 & \big\uparrow \\
 & \mathrm{SE}
\end{array}
$$

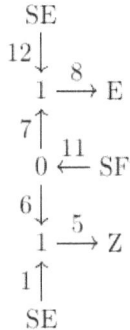

The sources of effort provide e_{12} and e_1, but these are both atmospheric pressure and so $e_{12} = e_1$. A consequence is that atmospheric pressure can be eliminated as a variable. This happens through the following set of relations:

$$e_6 + e_1 = e_5$$
$$e_1 = e_{12}$$
$$e_6 = e_7$$
$$e_7 + e_{12} = e_8 \text{ and so}$$
$$e_6 + e_1 = e_7 + e_{12} = e_8$$

Therefore, $e_5 = e_8$ and this effort is the sum of the atmospheric pressure and hose outlet pressure.

The sign convention of these pressures is not intuitive. We have modeled the hose as if it is pushing up on the bottom of the water in the bucket while the atmosphere pushes downward on the top of the water. If pushing downward is pushing in a positive direction, then pushing upward must be in a negative direction.

We have chosen to model the atmosphere as a positive pressure. Therefore, if flow comes from the hose outlet to push up on the water at the bottom of the bucket, then the outlet supplies this flow at a negative pressure. The sum of the positive atmospheric pressure and negative hose outlet pressure is all that we need to describe the dynamics of the system.

When we produce the system of equations for this graph and remove the redundant variables, the result is

$\dot{e}_8 = (f_{11} - f_5)/E$	Pressure at the bucket's bottom
$\dot{q}_8 = f_{11} - f_5$	Volume of fluid in the bucket
$f_5 = e_8/Z$	Flow through the pipe

We have included the volume of fluid in the bucket to show how the accumulated flow might be used even if it doesn't contribute to the dynamics of the system.

Water distribution system Let us construct a model of a water distribution network from several types of objects. A source of flow, such as a pump, provides water to a storage tank. A network of pipes carry water to points of use. Valves at those points control the flow at the ends of our water system.

To begin our modeling effort, consider a long and rigid section of pipe. The pipe section is free at both ends and so we leave these edges dangling. Within the pipe there is a mass of fluid and resistance to the flow of that fluid. This resistance might be caused by friction between the fluid and the wall of the pipe.

This implies two elements in the pipe: an F element and a Z element. We attach each element to a one junctions. Points of connection between the elements are modeled with zero junctions. Our convention for flow through the pipe will be from left to right. Hence, we arrive at the graph segment

$$\begin{array}{ccccccccccc} & & & & \mathrm{Z} & & & & \mathrm{F} & & \\ & & & & \uparrow & & & & \uparrow & & \\ \longrightarrow & 0 & \longrightarrow & & 1 & \longrightarrow & 0 & \longrightarrow & 1 & \longrightarrow & 0 \longrightarrow \end{array}$$

This graph segment can be reduced to its smaller equivalent

$$\begin{array}{ccc} & \mathrm{Z} & \\ & \uparrow & \\ \longrightarrow & 1 & \longrightarrow \\ & \downarrow & \\ & \mathrm{F} & \end{array}$$

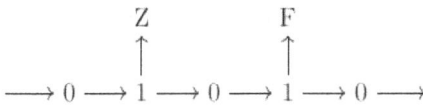

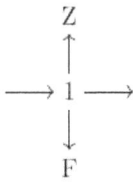

Using f for the single flow in the pipe, e_{in} for the effort on the left side of the pipe, and e_{out} on the right we see that

$$\dot{f} = (e_{\text{in}} - Zf - e_{\text{out}})/F$$

Another element in our water distribution system is the pump that supplies fluid. We can imagine a lake or river (a clean lake or river - we've left out the treatment plant!) that supplies the pump with water. This pump is operated in such a way that it maintains a particular quantity of water in a storage tower. An illustration of the pump in its house and the tower is shown in Figure 2.

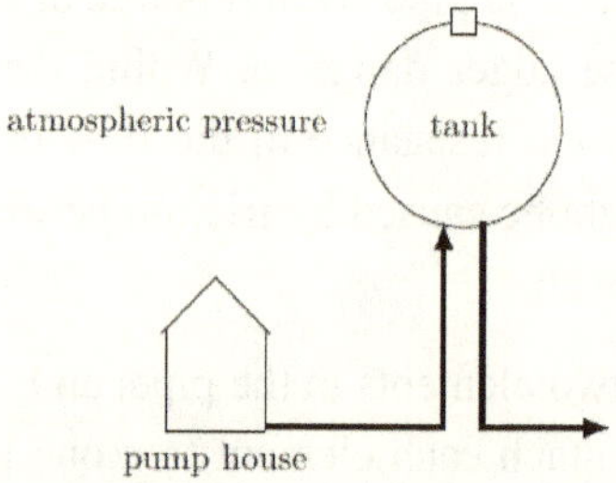

Figure 2. A water supply pump and storage tower.

This part of our system has two elements: the source of flow that models the pump and an E element that models the tank. The pump provides fluid that flows into the tank. The tank has a vent hole, shown at the top of our illustration, to keep the pressure on the surface of the stored fluid at atmospheric pressure. Therefore, the tank has two pressures: the pressure at its bottom where fluid is supplied to anything downstream and the pressure at the top into which fluid must be pumped. The graph segment for the tank and pump recalls the leaky bucket.

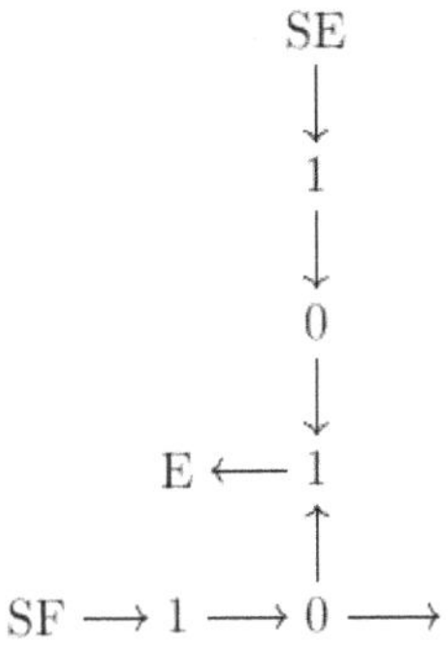

This graph can be simplified by eliminating redundant edges. We label the edges with descriptive text rather than numbers to more clearly label the variables in our system of equations. Doing so, we have

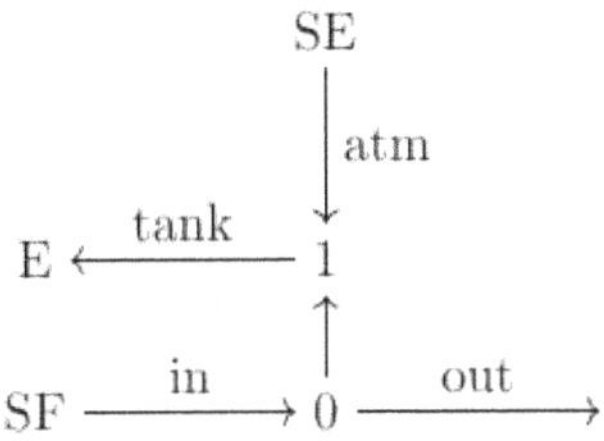

The system of equations is

$$\begin{aligned} \dot{e}_{\text{tank}} &= (f_{\text{in}} - f_{\text{out}})/E \\ e_{\text{out}} + e_{\text{atm}} &= e_{\text{tank}} \\ e_{\text{in}} &= e_{\text{out}} \end{aligned}$$

Lastly, we model the service points. Each service point comprises a Z element in series with atmospheric pressure. These service points model sinks, fountains, and other places where water leaves the network. The service point graph, after simplification, is

```
      Z
      ↑
 ⟶ 1 ← SE
```

Consider the water distribution system shown in Figure 3. This network has a pump and tank connected to two service points by three sections of pipe. To build a graph for this distribution network, we place a zero element wherever there is a connection between components. In place of one junctions we place the pipe, pump & tank, or service point in its appropriate place. The graph looks very much like the diagram, as we can see below.

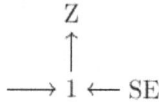

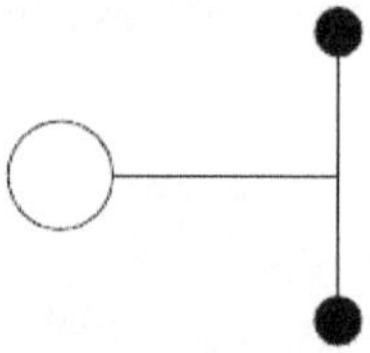

Figure 3. A water distribution network. The open circle is a pump house and tank. The closed circles are service points. The connecting lines are pipes.

```
                                     service
                                        ↑
                                        0
                                        ↑
                                       pipe
                                        ↑
pump & tank ⟶ 0 → pipe → 0
                                        ↓
                                       pipe
                                        ↓
                                        0
                                        ↓
                                     service
```

The graph can be completed by replacing the pump & tank, pipes, and service points with their graph fragments. We begin with the service points. To give unique labels to each Z element, we attach the subscripts 's1' and 's2' to resistance at the service points and 'p1', 'p2', and 'p3' to resistance in the pipes. The subscript 'atm' indicates atmospheric pressure.

$$
\begin{array}{c}
Z_{s1} \\
\uparrow \\
1 \longleftarrow SE_{atm} \\
\uparrow \\
0 \\
\uparrow \\
pipe \\
\uparrow \\
pump \& tank \longrightarrow 0 \rightarrow pipe \rightarrow 0 \\
\downarrow \\
pipe \\
\downarrow \\
0 \\
\downarrow \\
1 \longleftarrow SE_{atm} \\
\downarrow \\
Z_{s2}
\end{array}
$$

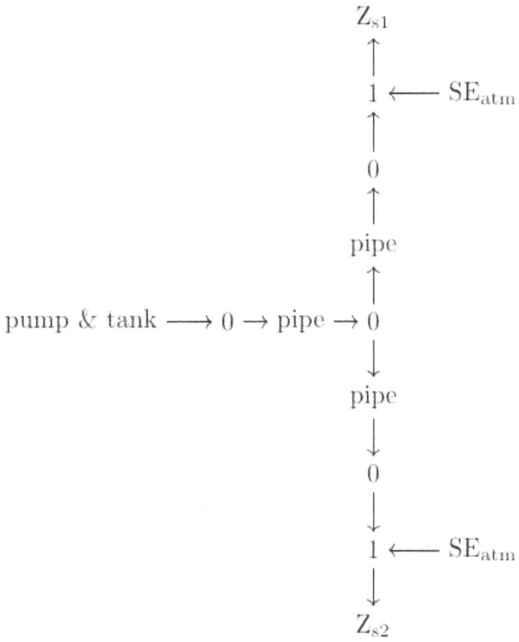

Continuing with the pipes and pump & tank elements we obtain the graph for the entire system in terms of its basic elements.

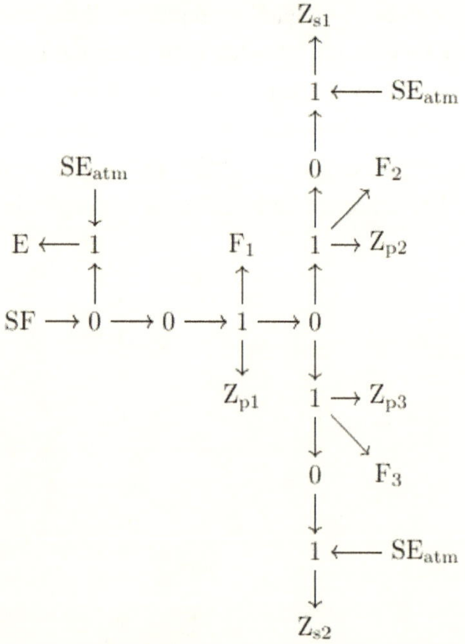

By removing redundant edges and replacing series Z elements with sum equivalents, we arrive at

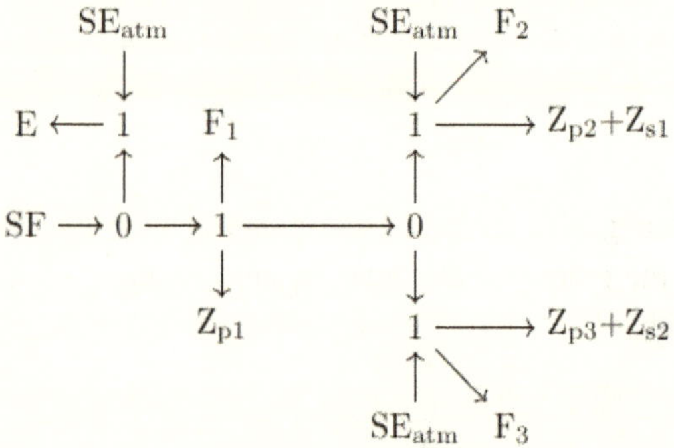

The graph can be simplified again if we measure all pressures as their difference from the atmospheric pressure. This allows us to treat atmospheric pressure as if it were zero.

That is, if we say that the pressure at a point is e then we mean that the total pressure at that point is equal to the atmospheric pressure plus e. This allows us to remove the sources of effort from our model.

Doing so and simplifying again, we obtain the following result. The edges are labeled so that we can extract a system of equations.

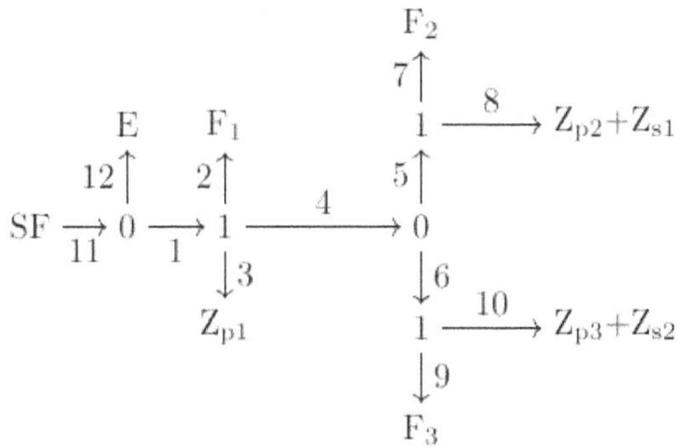

The pressure e_{12} at the bottom of the tank is

$$\dot{e}_{12} = (f_{11} - f_2)/E$$

where f_{11} is the flow from the pump. The flow out of the tank and down the supply pipe is

$$\dot{f}_2 = (e_{12} - f_2 Z_{p1} - e_4)/F_1$$

Examining the graph, we see that

$$f_2 = f_7 + f_9$$

The pressure e_4 at the junction of the pipes is

$$\begin{aligned} e_4 &= (Z_{p2} + Z_{s1})f_7 + \dot{f}_7 F_2 \\ &= (Z_{p3} + Z_{s2})f_9 + \dot{f}_9 F_3 \end{aligned}$$

The variables e_{12}, f_2, f_9, and f_7 are state variables obtained by (numerical) integration. The initial values for these quantities must be supplied, subject to the constraint on the sum of the flows. However, integration requires finding values for $\dot{e}_{12}$, $\dot{f}_2$, $\dot{f}_9$, $\dot{f}_7$, and e_4. These are five unknown variables for our five equations. This model has produced a (linear) differential-algebraic problem.

Home pipe system Figure 4 is a diagram of a pipe system in a home. It has a shower with a cold water valve, a hot water valve, and a valve that controls the flow of water out of the shower head. The sink has a cold water valve and a hot water valve. Cold water is supplied by a source at pressure P. Hot water is supplied by a source also at pressure P.

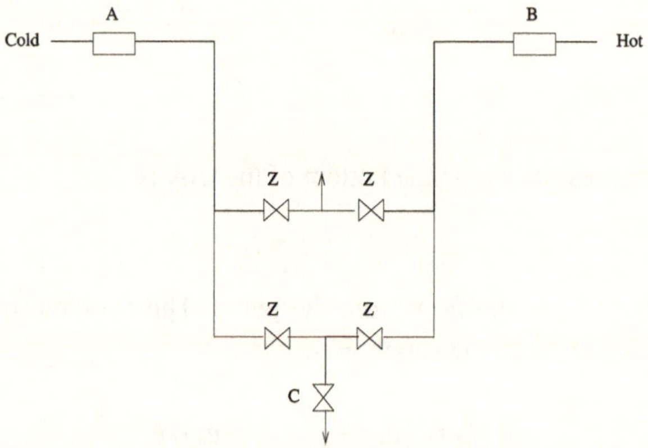

Figure 4. A home pipe system. The upper arrow shows flow out of the sink. The lower arrow indicates a shower head.

Water flowing from the cold source meets resistance Z_A. The intensity of resistance will depend on the layout of the piping, its diameter, length, and so forth. Water flowing from the hot source meets resistance Z_B. Cold and hot water valves are identical, each offering resistance Z to flow when the valve is open. The shower head offers resistance Z_C to flow when open. Closed valves prohibit flow.

Suppose that the cold water valve at the sink is open and all other valves are closed. The water flows from the cold water source to the sink, passing through the resistance Z_A of the cold water pipe and the resistance of the valve Z. The graph for this system is

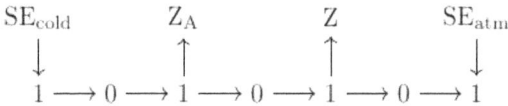

$$\begin{array}{ccccccccccccc} SE_{cold} & & & & Z_A & & & & Z & & & & SE_{atm} \\ \downarrow & & & & \uparrow & & & & \uparrow & & & & \downarrow \\ 1 & \longrightarrow & 0 & \longrightarrow & 1 & \longrightarrow & 0 & \longrightarrow & 1 & \longrightarrow & 0 & \longrightarrow & 1 \end{array}$$

We can see that there is a single flow f. If we define ΔP to be the difference between P and atmospheric pressure, then we can redraw the graph as

$$\begin{array}{ccccc} & & Z_A & & \\ & & \uparrow & & \\ SE & \longrightarrow & 1 & \longrightarrow & Z \end{array}$$

The equation for this graph is

$$\Delta P = (Z_A + Z)f$$

On a busy morning (or evening, or afternoon, as your busy life may dictate!) all of the valves can be open. This causes warm water to emerge from the sink and shower, the warm water being a mixture of the hot and cold flows. The graph for this system of open valves mimics the diagram. Sources of effort model the cold and hot water sources, and each of the flows emerges into a source of effort that is the atmospheric pressure.

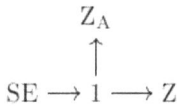

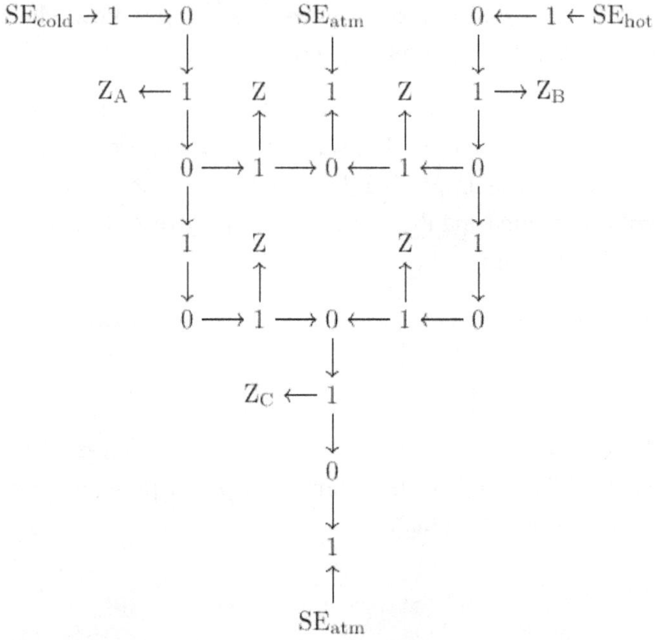

By collecting junctions that merely introduce equalities, this graph can be reduced as shown below.

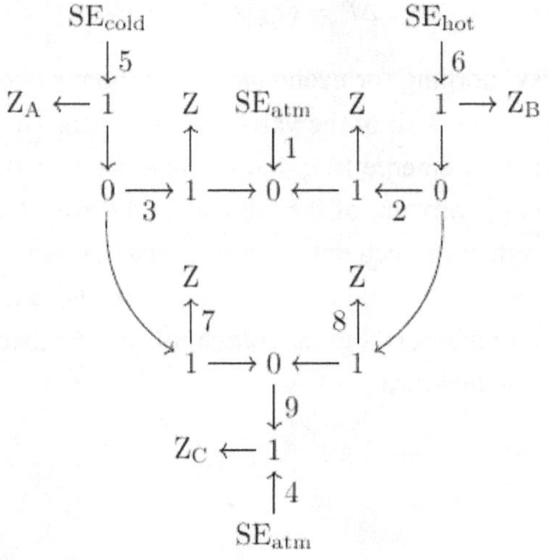

This graph has no dynamic behavior. The flows and efforts have a simple linear relationship. There are seven distinct flows: f_5 emerging from the cold water source and f_6 from the hot water source; f_3 through the cold water valve of the sink and f_2 through the hot water valve; f_7 through the cold water valve of the shower and f_8 through its hot water valve; and f_9 through the flow control valve of the shower. The flow f_1 out of the sink is $-(f_2 + f_3)$.

Likewise, there are seven distinct efforts: e_5 and e_6 at the sources; e_3 and e_2 at the junctions where flows split between the sink and shower pipes; e_9 at the junction of the shower head; and e_4 and e_1 at the exits of the pipe system. The seven flows and seven efforts are related by

$$
\begin{aligned}
e_5 - e_3 &= f_5 Z_A \\
e_6 - e_2 &= f_6 Z_B \\
e_3 - e_1 &= f_3 Z \\
e_2 - e_1 &= f_2 Z \\
e_3 - e_9 &= f_7 Z \\
e_2 - e_8 &= f_8 Z \\
e_9 + e_4 &= f_9 Z_C
\end{aligned}
$$

Constant power pump A long section of hose contains a mass of fluid. Friction in the hose creates a resistance to flow. A pump that supplies constant power pushes fluid through the hose. Fluid flows out of the hose at atmospheric pressure. This system is illustrated in Figure 5.

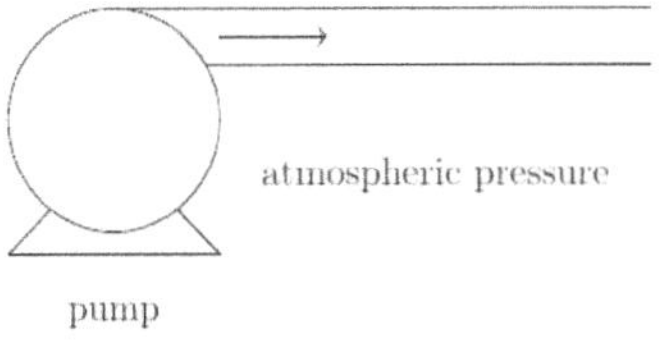

Figure 5. A pump at one end of a length of hose.

This model has four elements. An F element for the fluid mass, a Z element for the resistance to flow, a source of effort supplies atmospheric pressure, and a source of power models the pump. Each element is adjacent to a one junction. The elements are connected by a long hose, which introduces zero junctions between the one junctions of the elements.

$$\begin{array}{cccc}
\text{SP} & \text{F} & \text{Z} & \text{SE} \\
\downarrow & \uparrow & \uparrow & \downarrow
\end{array}$$

$$1 \longrightarrow 0 \longrightarrow 1 \longrightarrow 0 \longrightarrow 1 \longrightarrow 0 \longrightarrow 1$$

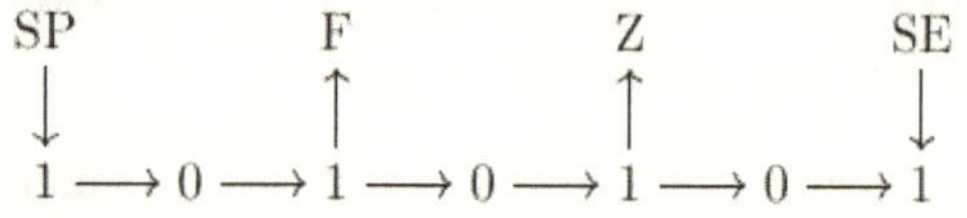

This graph can be reduced to a single one junction with the four elements attached to it.

$$\begin{array}{c}
\text{F} \\
\uparrow 3 \\
\text{SP} \xrightarrow{\;1\;} 1 \xleftarrow[2]{} \text{SE} \\
4 \downarrow \\
\text{Z}
\end{array}$$

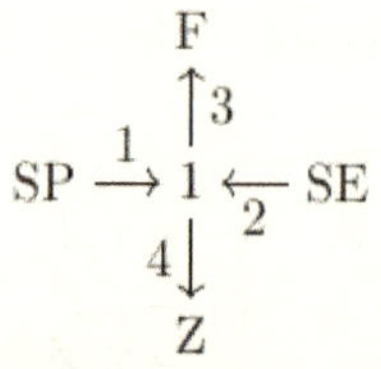

The model has a single flow f and four distinct efforts. The equations produced by this graph are

$$\dot{f} = e_3/F$$
$$e_4 = Zf$$
$$e_1 + e_2 = e_3 + e_4$$
$$P = e_1 f$$

The variables P and e_2 are given by the source elements. There are four unknowns f, e_3, e_1, and e_4 and four equations. Unlike our previous examples, this model contains the nonlinear product $e_1 f$ that was contributed by the source of power. Consequently, $f \neq 0$ and $e_1 \neq 0$ are required if $P \neq 0$.

If we know the rate of flow that is desired and the resistance Z, then this model can be used to calculate the power that is needed to supply this rate of flow in a steady state of operation. The steady state of a system is where its time derivatives are zero. Substituting zero for $\dot{f}$, we find $e_3 = 0$ and

$$P = (Zf - e_2)f$$

Exercises

Ex. 4.1 Show that a section of pipe with length ℓ, cross-sectional area A, and containing a fluid of density ρ can be modeled with an F element for which $F = \rho\ell/A$.

Ex. 4.2 A pump acting as a source of effort draws contaminated water from a reservoir that is open to the air. The water is pushed by the pump through a long pipe that ends in a sand filter. The pipe contains a substantial mass of fluid but offers no resistance to flow. The filter offers a constant resistance to the flow through it. Water emerges from the filter into a reservoir of clean water open to the air. Draw a graph of this system and write its equations. Hint: the problem is simpler if you measure pressure as the difference from atmospheric pressure.

Ex. 4.3 Create a graph for a water distribution system like the example in this chapter. However, in this case the pump is attached directly to a long line. At the end of this line, the flow is diverted left into a storage tank and right to a service point. Write the equations for this graph. Measure pressure in relation to the atmosphere. Suppose the pump fails and merely provides atmospheric pressure. Write the graph and system of equations after this event.

Ex. 4.4 Consider the piping diagram shown in Figure 4. Draw the graph and derive equations for flow given the following configurations of the valves: (1) shower cold valve and C valve open, all other closed, (2) C closed and all others open.

Ex. 4.5 In the piping diagram of Figure 4, the flow f_4 is negative when water comes out of the shower. The flow f_1 is negative when water comes out of the sink. Redraw the graph so that the flow variables representing water from the sink and shower are positive when water is flowing.

Ex. 4.6 Draw a graph and derive equations for the system shown below. Assume the tanks are closed to the atmosphere. The supply pipes provide constant flow. The pump extracts fluid to maintain constant pressure. Create two models: (a) the 'x' elements are resistance to flow (e.g., partially closed valves); (b) the 'x' elements are long lines with inertia and resistance to flow.

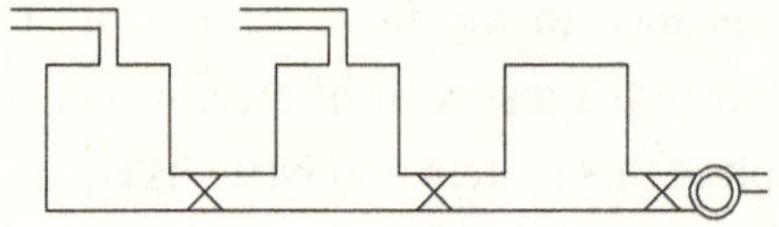

5. Mechanics

Mechanics is the study of forces acting upon physical objects and of the motions of those objects as you would ordinarily encounter them. It is "the branch of applied mathematics dealing with motion and forces producing motion."[4] If you push a plate across a table, compress the spring in a retractable pen, or build a trebuchet, then you are working with mechanics.

We will study two types of mechanical motion: translating and rotating. Transformers are used to incorporate both into a single mechanical machine. Flow in a translating machine is velocity, which is measured in meters per second. Effort in a translating machine is force measured in Newtons. The product of these is power.

The effort in a rotating machine is torque, which has units of Newtons times meters (N·m). We have several options for flow, which differ from each other by a constant factor. To see that these choices result in power, let us use rotations per second, or Hertz (Hz), for flow. This has units of 1/s. Multiplying effort and flow we get

$$ef = \mathrm{N} \cdot \mathrm{m} \cdot \frac{1}{\mathrm{s}} = \frac{\mathrm{J}}{\mathrm{s}} = \mathrm{W}$$

Our choice of flow will result from our choice for measuring distance. Motion is circular in a rotating machine, and so we may use degrees, radians, or rotations for this measure. Radians are common in practice, and so we will use units of radians per second. One rotation of the machine passes it

[4] From Oxford Languages online dictionary, accessed February 2025.

through 2π radians. Therefore, a rotational speed of 1 Hz is equal to 2π radians per second.

0 and 1 junctions The one junction in a mechanical system is a point where two or more components are rigidly connected. A rigid connection cannot be deformed, and the attached elements move together. Consequently, a rigid connection imposes an equal velocity on each of its attached elements.

Our idealized point of connection, being without mass and rigid, cannot have work done on it. Therefore, the power transmitted to the point of connection must leave it. There is a single flow f at the junction. The forces e_1, e_2, ..., e_n acting upon the connection must be such that the total power satisfies

$$fe_1 + fe_2 + \cdots + fe_n = 0$$

Dividing by the flow, we find that the sum of the efforts into the one junction must be zero.

A zero junction models the force or torque that bends, stretches, twists, slides, and otherwise works on an object. The zero junction, like the one junction, does not store or dissipate power. Therefore, the flows must be such that

$$ef_1 + ef_2 + \cdots + ef_n = 0$$

Dividing by the single effort, we find that the sum of the flows must be equal to zero.

I find the mechanical interpretation of zero and one junctions to be less intuitive than in other types of system. Very likely, this is because of my (long ago) undergraduate training in the analysis of electric circuits, which happen to look very much like hydraulic systems and very unlike mechanical systems. Old habits die hard. What we will find is that one junctions model the velocities of masses and zero junctions model the forces that compress and stretch springs or that are dissipated by friction.

Basic elements The Z element in a mechanical system typically represents friction, which converts the energy of a moving object into heat. The E element stores energy by deforming and returns energy by resuming its normal shape. A spring is the canonical E element. The F element is inertia, which is mass in a translating system and angular mass in a rotating system.[5]

The laws of Newtonian mechanics are evident in the equations imposed by the F element. First, force equals mass times acceleration. The equation for the F element is

$$e = F\dot{f}$$

where e is force, F is mass, and $\dot{f}$ is the derivative of velocity, which is acceleration.

In a rotating system, calculations of mass involve complex considerations of geometry. Regardless, the F element retains its role as a mass, in this case the angular mass with units of kg·m^2. The product of angular mass and rotational acceleration is torque, which has units of

$$\mathrm{kg} \cdot \mathrm{m}^2 \cdot \frac{1}{\mathrm{s}^2} = \mathrm{N} \cdot \mathrm{m}$$

[5] Angular mass may also be called moment of inertia, mass moment of inertia, rotational inertia, or rotational mass according to the author's preference.

This is the rotating analog of force equals mass times acceleration.

The momentum of a moving body is the product of its mass and velocity. Integrating both sides of the equation for an F element, we find that

$$\int e\,dt = F\int \dot{f}\,dt$$

and so

$$p = Ff$$

Hence, accumulated effort has the physical interpretation of momentum.

Hooke's law of springs is evident in the equation for the E element. Hooke's law states that a spring with compliance E, measured in displacement per unit force, requires a force e to compress (or extend) the spring through a distance q according to

$$e = q/E$$

Taking the derivatives of both sides we arrive at

$$\dot{e} = f/E$$

In a rotational system, energy is stored by twisting the spring rather than by stretching or compressing. The units of the spring in this case are radians per unit torque. The physical principle is the same.

A note of caution! The units of our E element describe *compliance*. Mechanical engineers more often use *stiffness* to describe a spring or spring like element. The stiffness $1/E$ has units of force per unit displacement. The symbol K is commonly used for stiffness, whereas we will use the lower case k for compliance.

In summary, we have the following mechanical properties described by the Z, E, and F elements.

1. Friction is modeled with a Z element.
2. Mass and angular mass are modeled with an F element.
3. A spring or similar element that stores energy by deformation is modeled with an E element.

Transformers In a rotating system, a gear or a collection of gears is used to trade rotational speed for torque. To understand the function of a gear, imagine two interlocking wheels such as those shown in Figure 6. To discuss gears, we use the angle θ for the accumulated flow q and the rotational speed ω for the flow f.

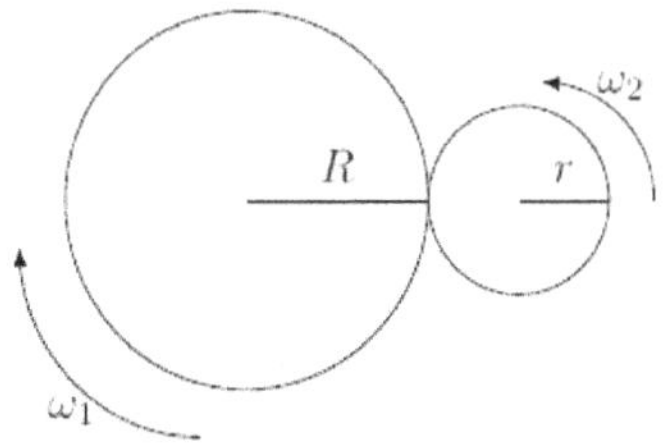

Figure 6. Two interlocking wheels.

The large wheel of radius R turns at speed ω_1. Suppose that we mark a point on the rim of the large wheel and then turn it through an angle of θ_1 radians. The marked point will travel a distance $\theta_1 R$ and in 2π radians it travels the circumference of the circle.

Suppose that we also mark a point on the rim of the small wheel. The small wheel, because it is interlocked with the large wheel, must rotate so that the point on the small wheel moves the same distance as the point on the large wheel. Therefore, geometry requires

$$\theta_1 R = \theta_2 r$$

Define

$$n = R/r$$

so that

$$\theta_1 n = \theta_2$$

The number n is the gear ratio, which is the number of times the driven gear (the small wheel in this example) turns for each turn of the driving gear (the large wheel in this example). Taking the derivative with respect to time of both sides we find

$$\omega_1 n = \omega_2$$

This is the equation of a transformer with constant of transformation $T = n$ if our convention is that the speed of the driving gear points into the transformer element.

In practice, we must be careful to check our use of a supplied gear ratio to be sure which is the driving gear and which is the driven gear. We must also pay attention to relative motions. In our specific case, the directions of rotation of the two wheels are in opposition. If the large wheel rotates clockwise, then the small wheel rotates counterclockwise.

If energy is not stored or dissipated at the point of contact between the wheels, then the transformation of speeds must occur without loss of power. If τ_1 is the torque applied to the large wheel and τ_2 the torque applied to the small wheel then conservation of power requires

$$\omega_1\tau_1 = \omega_2\tau_2 = n\omega_1\tau_2$$

and so

$$\tau_1 = n\tau_2$$

This is the relation of efforts at a transformer with $T = n$.

A rack and pinion, like the one shown in Figure 7, interchanges rotational and translational motion. This device comprises a wheel and rod that are interlocked. The wheel of radius R rotates at ω radians per second. The rod translates at the velocity of the wheel's perimeter.

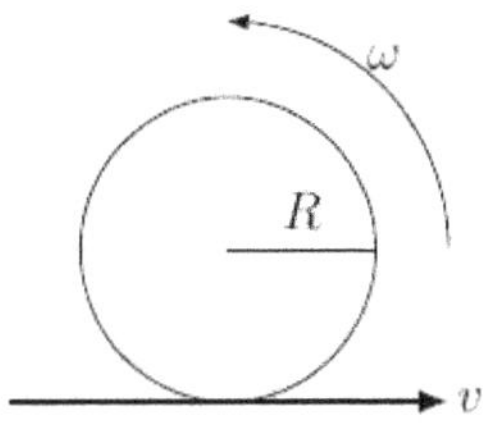

Figure 7. Rack and pinion.

A complete turn of the wheel moves the rod a distance of $2\pi R$. The time required to rotate through 2π radians is $(2\pi)/\omega$ seconds. Therefore, the velocity v of the rod is

$$v = 2\pi R\frac{\omega}{2\pi} = R\omega$$

If we relabel this expression using flows so that $v = f_1$ and $\omega = f_2$ then $1/R$ becomes the constant T of a transformer; that is

$$f_1/R = f_2$$

The transformer's expression for effort follows from conservation of power.

Drawings to graphs To create a graph, we must choose the direction of positive motion. The choice is arbitrary. However, consistency will simplify the system of equations.

In translating systems, a convenient convention is to have positive motion coincide with the typical orientation of a line chart. Positive motion moves the system within the drawing from left to right and from bottom to top, just as we would draw a chart with the positive x-axis going from left to right and the positive y-axis from bottom to top.

In rotating systems, positive motion can be in the natural or preferred direction of rotation. For instance, if the rotation occurs along with translation, then a natural direction of rotation coincides with positive translation when a transformer is introduced. For interlocked wheels, the natural direction might be dictated by the driving gear.

Often, the preferred direction for effort comes from the graph as you construct it. However, when defining sources of effort, we are generally required to make deliberate choices. Our convention for orientating efforts is that positive effort creates positive motion; negative effort creates negative motion.

For example, the engine of a locomotive supplies a positive force when pushing the train from left to right, thereby imparting a positive velocity in the x direction. It supplies a negative force when pushing the train from right to left, thereby imparting a negative velocity in the x direction.

With directions for efforts and flows in hand, we look at the system's components. We attach each F element (that is, each mass) to a one junction. Each E and Z element is attached to a zero junction. These elements exert effort in response to relative motion across the element. Attach each source of effort to a zero junction and each source of velocity to a one junction.

Next, place a one junction at each point where elements are connected. Draw edges between these connecting one junctions. A helpful organizing principle is, if you can, to orient these edges coincident with the direction of positive motion in the drawing. Several examples will illustrate the procedure.

Mass, spring, and damper Figure 8 shows a mass connected to a spring and damper. A force acts on the mass from the left. The spring and damper are connected to an immobile wall on the right. This system has five components: an F element, which is the mass; a source of effort pushing on the mass; an E element, which is the spring; a Z element, which is the damper; and a source of velocity, which is the immobile wall.

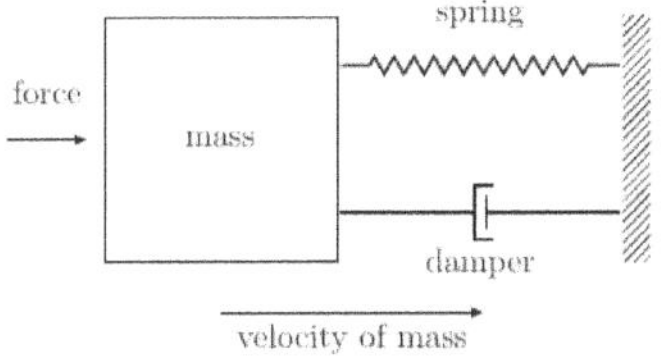

Figure 8. A mass, spring, and damper.

The incomplete graph below shows the model elements and their adjacent junctions. The junctions without attached elements are points of connection in the drawing. There are three of these points: where the force connects to the mass; where the mass, the damper, and the spring are connected together; and where the damper, the spring, and the wall are connected to each other.

$$
\begin{array}{ccccccc}
 & & & & Z & & \\
 & & & & \uparrow & & \\
 & & F & & 0 & & \\
 & & \uparrow & & & & \\
SE \longrightarrow 0 & 1 & 1 & 1 & & 1 & 1 \longleftarrow SF \\
 & & & & & & \\
 & & & & 0 & & \\
 & & & & \downarrow & & \\
 & & & & E & &
\end{array}
$$

In the diagram, we have drawn an arrow to show that the direction of positive motion is from left to right. The missing edges in the graph are oriented accordingly. The zero junction attaching the source of force to our graph has its edge pointing towards the neighboring one junction. If a positive effort is imposed on the zero junction and there is positive velocity then power flows from the source to the mass.

The orientations of the remaining edges are chosen to be consistent with this first choice. Conveniently, our graph has a near one to one correspondence with the drawing, including the placement of elements and the orientation of edges in the direction of positive flow. We omit labels for the moment to avoid cluttering the graph as we simplify it.

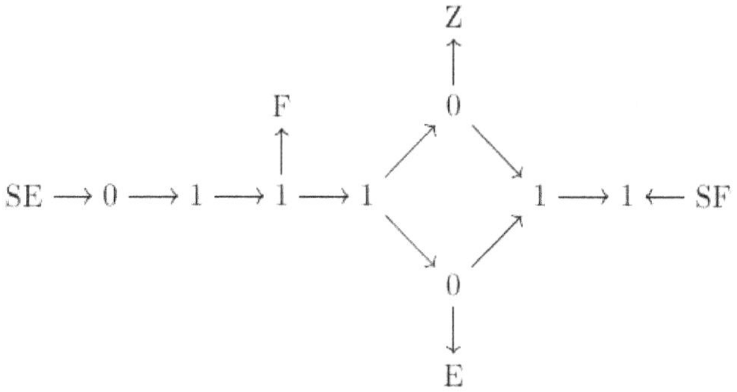

We can reduce the groups of adjacent one junctions to single junctions.

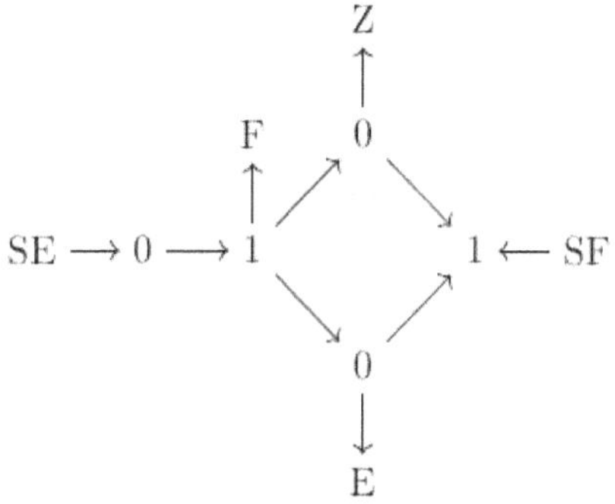

Two further simplifications are possible. First, the zero junction connecting the source of effort to the one junction of the mass is redundant. These edges have equal effort and equal flow. Second, the diamond can be replaced with a simpler, equivalent structure.

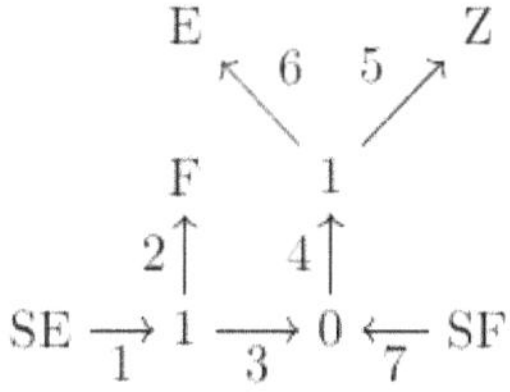

Because the wall is immobile, the flow $f_7 = 0$. It follows that $f_3 = f_4$. This allows us to simplify the graph again.

$$
\begin{array}{ccc}
 & F & \\
 & 2\uparrow & \\
\text{SE} \xrightarrow{\ 1\ } & 1 & \xrightarrow{\ 5\ } Z \\
 & \downarrow 6 & \\
 & E &
\end{array}
$$

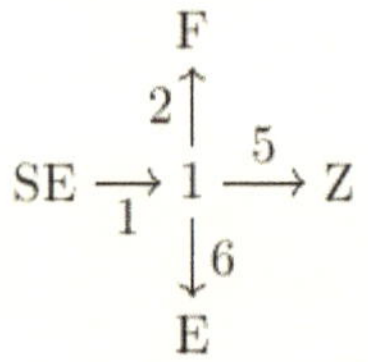

The equations of motion for this graph are

$$\dot{f}_2 = \frac{1}{F}\left(e_1 - e_6 - Zf_2\right)$$
$$\dot{e}_6 = \frac{1}{E}f_2$$

Vehicle suspension The suspension in a vehicle absorbs changes in the level of a road surface to reduce pitching and heaving of the passenger compartment. The suspension also keeps your vehicle's tires in contact with the road when it encounters a significant bump. Here we assume that the tires stay in contact with the road.

A suspension system connecting a single tire to a vehicle has four mechanical parts: the mass of the vehicle that rests on the suspension, the suspension itself consisting of a spring in parallel with a damper, the mass of the tire, and the springiness of the tire's inflated, rubbery material. The suspension system is illustrated in Figure 9.

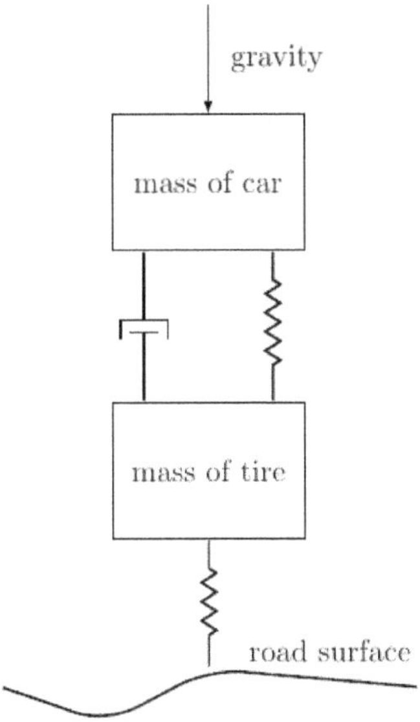

Figure 9. Tire, suspension, and car body.

A source of flow and a source of effort act on the system. The source of flow models the changing height of the road surface as the wheel rolls forward. A vehicle traveling at constant speed V moves a distance Vt in t units of time. If we know the road height h as a function of distance, then the source of velocity v applied to the tire is

$$v(t) = \dot{h}(Vt)$$

Because h has units of distance (meters above or below some reference road surface), the derivative of h with respect to time is velocity (change in meters above or below the reference road surface per unit time), and hence our source of velocity acting on the tire.

The source of effort is the force produced by gravity acting on the body of the car. Of course, gravity acts on all parts of the suspension. However, the tire, which we have assumed stays in contact with the road and whose motion on that surface has been modeled as a source of velocity, cannot be accelerated downward through the impeding Earth. For the other elements, we assume that their masses are much smaller than the mass of the car body and can be ignored.

On Earth, the acceleration due to gravity g is 9.8 $\mathrm{m/s^2}$. A mass of m kg experiencing this gravity exerts a force equal to

$$-mg = \mathrm{kg} \cdot \frac{\mathrm{m}}{\mathrm{s}^2} = \mathrm{N}$$

This is the weight of the object. The negative sign is consistent with our convention that positive power coincides with motion in the direction of the force, that is, in the direction of decreasing y.

In this example, we will augment the notation for our E, F, SE, and SF elements with a colon followed by a symbol for the element's parameter. For instance, the SF element modeling change in the height of the road becomes SF: v. We use m_1 for the mass of the car so that it appears as F: m_1 in the graph. The SE element that models gravity is SE: $-m_1 g$. Likewise for the tire with mass m_2, the spring constant k_1 of the suspension, and the springiness of the tire k_2. The damper has a damping coefficient b.

We begin our modeling effort by reusing the graph of the mass, spring, and damper system from the previous example. Looking back at the drawing for that model, it is almost exactly the suspension and mass of the car body. To reuse this graph, we remove the SF element that models the immobile wall, leaving that end dangling. To avoid cluttering the graph, we omit the edge labels until they are needed to extract a system of equations.

$$\text{E: } k_1 \qquad\qquad \text{Z: } b$$
$$\nwarrow \qquad \nearrow$$
$$\text{F: } m_1 \qquad 1$$
$$\uparrow \qquad \uparrow$$
$$\text{SE: } -m_1 g \longrightarrow 1 \longrightarrow 0 \longleftarrow$$

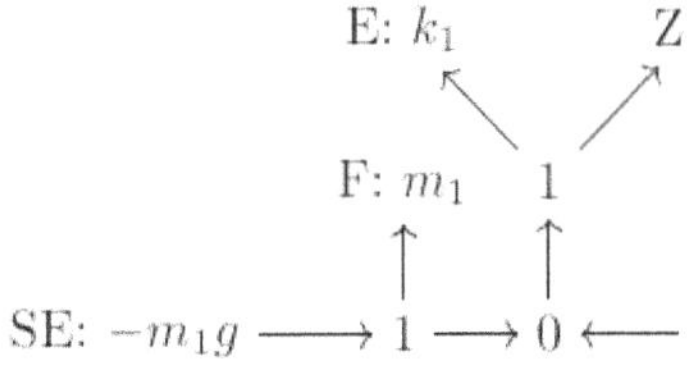

Leaving this to the side, we build a graph that models the tire and the road. This graph has a one junction where the mass of the tire is connected to the suspension; a one junction for the mass of the tire; a one junction where the mass of the tire connects to the spring; a zero junction for the spring; a one junction where the spring connects to the road; and a one junction for the road. We leave dangling the one junction that will connect this graph fragment to the graph fragment above.

$$\text{F: } m_2 \qquad\qquad \text{E: } k_2$$
$$\uparrow \qquad\qquad\quad \uparrow$$
$$\longrightarrow 1 \longrightarrow 1 \longrightarrow 1 \longrightarrow 0 \longrightarrow 1 \longrightarrow 1 \longleftarrow \text{SF: } v$$

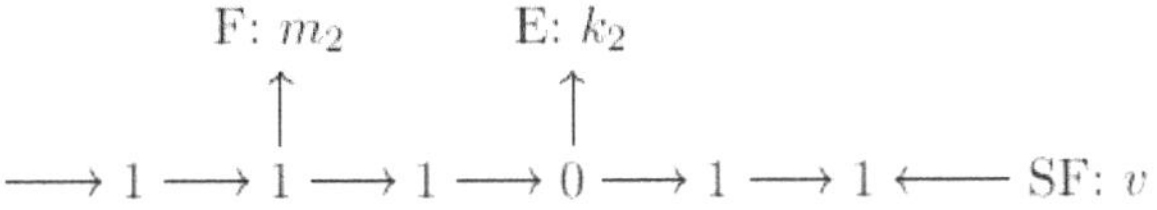

This graph fragment can be simplified by combining adjacent one junctions.

$$\text{F: } m_2 \quad \text{E: } k_2$$
$$\uparrow \qquad \uparrow$$
$$\longrightarrow 1 \longrightarrow 0 \longrightarrow 1 \longleftarrow \text{SF: } v$$

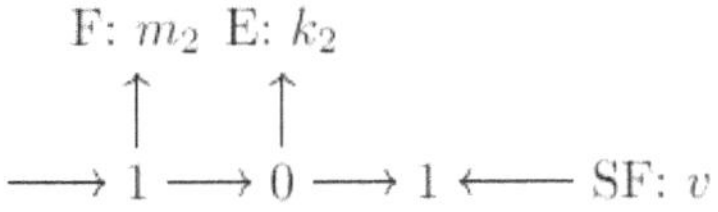

The one junction adjacent to the SF element has both edges pointing inward. If we keep this arrangement, then the effort of the edge adjacent to the SF element has the opposite sign of the effort imposed by the E element, but these efforts are equal in magnitude. Hence, the effort adjacent to the SF element is redundant. It merely introduces an equality statement into our system of equations. This one junction can be eliminated to get the graph fragment below.

$$\text{F: } m_2 \quad \text{E: } k_2$$
$$\uparrow \qquad \uparrow$$
$$\longrightarrow 1 \longrightarrow 0 \longleftarrow \text{SF: } v$$

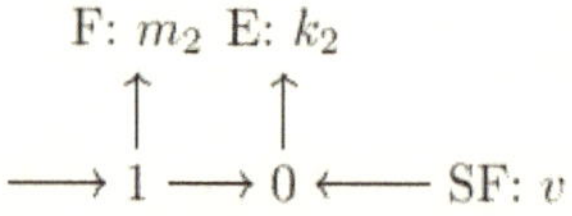

The graph fragments for the mass, spring, and damper system and for the tire and road are combined with a one junction at the point of connection. Where the two graphs meet at this connecting junction, the directions of the edges introduce an inconvenient sign convention.

If, for the moment, we label the edge on the left of the one junction 1 and the edge on the right 2, then the outward pointing edges imply $e_1 = -e_2$. The edges that point to the zero junction inherited from the mass, spring, and damper system present a similar inconvenience.

$$\text{E: } k_1 \qquad\qquad \text{Z: } b$$
$$\nwarrow \qquad \nearrow$$
$$\text{F: } m_1 \quad 1 \qquad\qquad \text{F: } m_2 \ \text{E: } k_2$$
$$\uparrow \quad \ 3\uparrow \qquad\qquad\quad \uparrow \qquad \uparrow$$
$$\text{SE: } -m_1 g \longrightarrow 1 \xrightarrow{4} 0 \xleftarrow[1]{} 1 \xrightarrow{2} 1 \longrightarrow 0 \longleftarrow \text{SF: } v$$

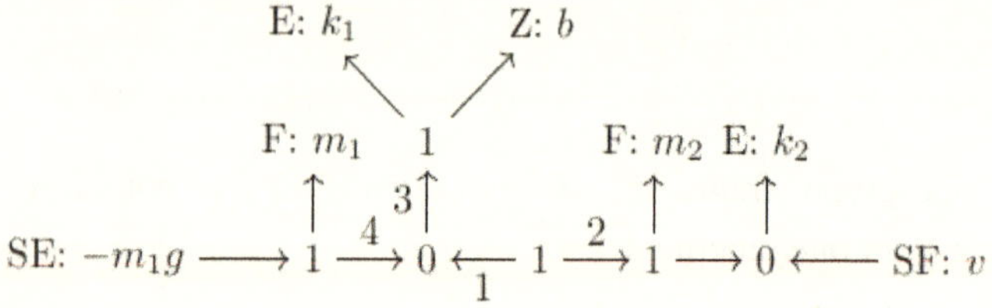

However, we are free to choose the directions of edges 1 and 2! Let us choose that positive velocity moves the body of the car upward, away from the road surface. With this choice, the road supplies positive power when it is rising and pushing the tire up. Using this upward convention for motion, we point the edges with directions that we are free to choose in the same direction as the source of flow; that is, toward the mass of the car body.

What of the edges labeled 3 and 4 in the diagram? We may also choose their directions. The k_1 element stores power and the b element dissipates power. Hence, the power entering the zero junction from the right (the edge labeled 1), which originates with the motion of the road, will be split into two parts.

Part of the power flows along edge 4 and into the car body. Another part of the power flows along edge 3 and into the damper and spring. Hence, edge 4 should point from the zero junction into the one junction to align with the direction of positive power.

Shown below is our finished rendition of the graph. The edges have been reoriented as described above and the adjacent one junctions combined into a single one junction. The edges are relabeled so that elements with a state variable are adjacent to a numbered edge.

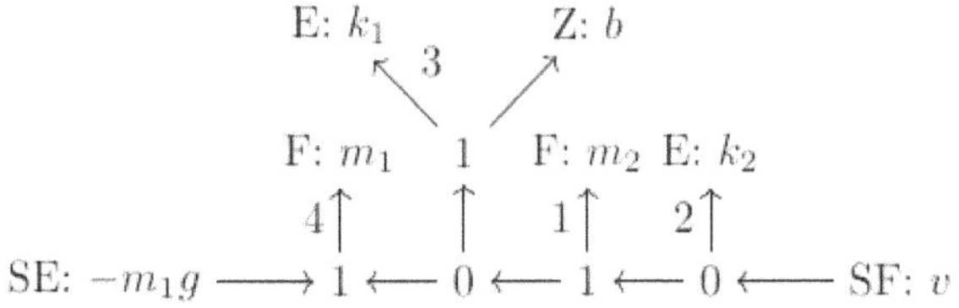

Reading off the system of equations and simplifying, we find

$$
\begin{aligned}
\dot{e}_2 &= (v - f_1)/k_2 \\
\dot{e}_3 &= f_3/k_1 \\
\dot{f}_1 &= (e_2 - e_3 - f_3 b)/m_2 \\
f_3 &= f_1 - f_4 \\
\dot{f}_4 &= (e_3 + f_3 b - m_1 g)/m_1
\end{aligned}
$$

Mechanical lift A mechanical lift is illustrated in Figure 10. A torque τ is applied to Gear #1. This gear has an angular mass J_1 and a radius R_1. Gear #1 drives Gear #2, which has an angular mass J_2 and a radius R_2. Gear #2 is connected to the wheel by a shaft that can be deformed. The shaft has rotational compliance k_1 and its bearing has friction b.

Around the wheel coils a cable that stretches under the weight of the load. This cable acts like a spring with compliance k_2. The load is a mass m. Gravity pulls the mass toward Earth.

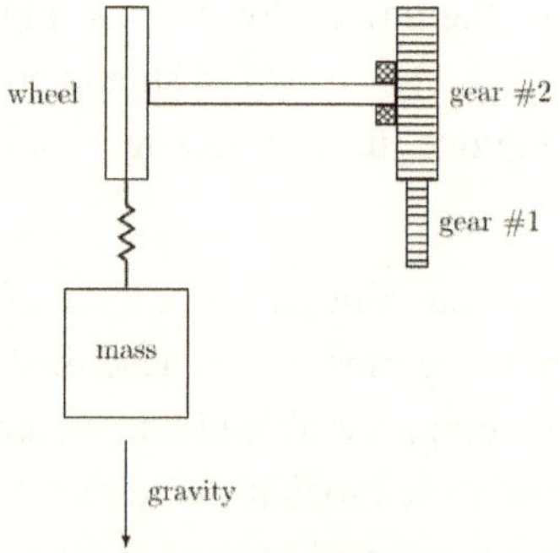

Figure 10. The components of a mechanical lift.

We can assemble a model of the lift from pieces. Our first piece comprises Gear #1, Gear #2, and the shaft. There are four points of connection in this assembly. First, where the torque attaches to Gear #1. Second, where Gear #1 attaches to the driving side of a transformer that models the interlocking of the gears. Third, where Gear #2 attaches to the driven side of the transformer. The transformer has a transformation coefficient $T_1 = R_2/R_1$.

The final connection is Gear #2 to the shaft. The shaft is modeled with two components in parallel: the shaft compliance and the bearing friction. For the moment, the shaft's end is left dangling. We choose the edge directions so that power flows from the source of effort, through the components of the system, and then out of the graph fragment via the shaft. Our procedure for producing the graph yields the following result.

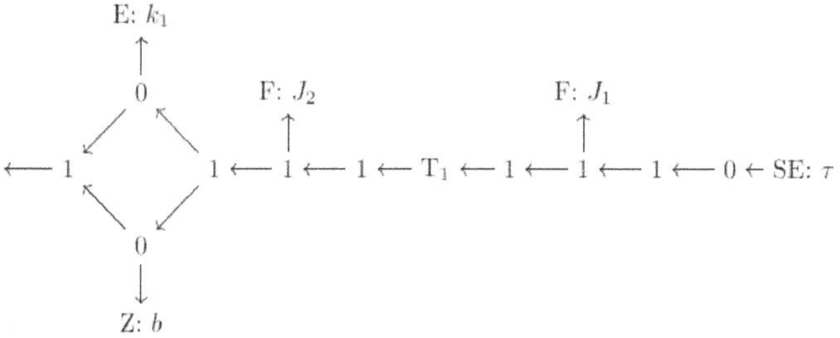

The graph can be simplified by combining adjacent one junctions, eliminating junctions with degree two, and replacing the diamond with an equivalent structure.

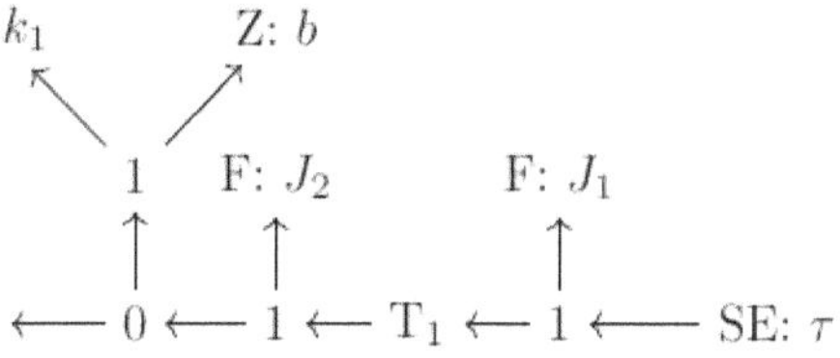

At the end of the shaft we connect the wheel, which has an angular mass of J_3 and a radius R_3. The wheel and cable act as a rack and pinion with constant of transformation $T_2 = R_3$. The springiness of the cable is connected, at one end, to this transformer and, at the other, to the mass m. Lastly, the mass is connected to a source of effort, which is the force of gravity $-mg$. The graph fragment for this part of this system, after simplification, is shown below.

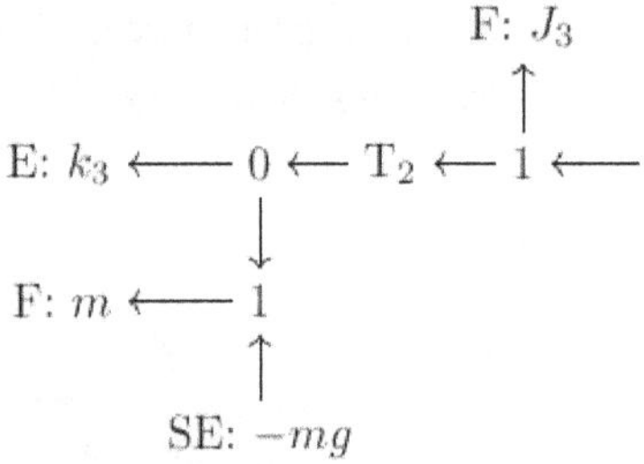

Connecting the fragments through a one junction, we arrive at the complete model. We could, if we wished, simplify this model by eliminating the connected one junctions. The orientation of the edges has power flowing through the system from the applied torque to the mass.

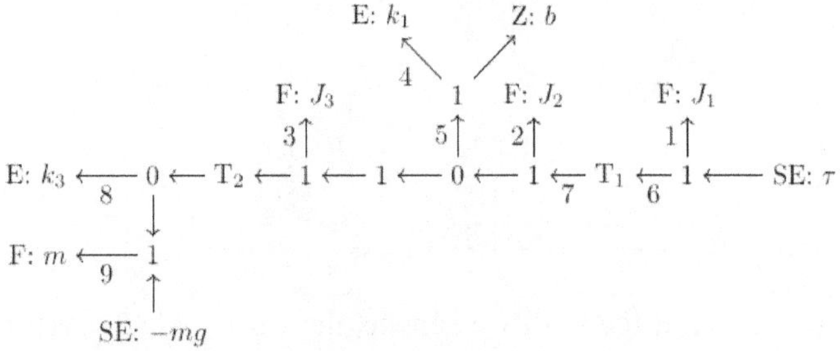

The system of equations for this graph is a set differential algebraic equation, where the algebraic constraints originate from the transformer that connects our idealized gears. After some simplifying of the equations, we find

$$J_1\dot{f}_1 = \tau - e_6$$
$$J_2\dot{f}_2 = e_7 - e_5$$
$$T_1 e_7 = e_6$$
$$T_1 f_1 = f_2 \text{ so that } T_1\dot{f}_1 = \dot{f}_2$$
$$k_1\dot{e}_4 = f_5$$
$$f_5 = f_2 - f_3$$
$$e_5 = e_4 + f_5 b$$
$$J_3\dot{f}_3 = e_5 - e_8/T_2$$
$$k_3\dot{e}_8 = T_2 f_3 - f_9$$
$$m\dot{f}_9 = e_8 - mg$$

At each step of the (numerical) integration procedure, the known variables are f_1, f_2, e_4, τ, f_3, e_8, and f_9. The unknowns are e_6, e_7, e_5, f_5, and the six derivatives with respect to time. We have ten unknown variables in ten independent equations, a solvable system.

The direction of rotation is not obvious from the illustration. One possibility is that a positive torque acting on Gear #1 turns that gear in the counterclockwise direction. This causes Gear #2 and the wheel to turn clockwise. If the cable is wound so that clockwise motion of the wheel raises the mass, then gravity imparts positive power when the mass is falling (it accelerates the mass toward Earth) and negative power when the mass is rising (it impedes upward motion).

Constant power motor A direct analog to the constant power pump from our hydraulic examples is a rocket of mass m being pushed up through the air. The air opposes this motion with resistance Z and gravity imparts a downward force. The graph is identical to the hydraulic case: a one junction connecting the elements F: m, Z, SP, and SE: $-mg$. The power required to propel the rocket upwards at a steady state speed f is

$$P = (Zf + mg)f$$

Exercises

Ex. 5.1 Derive the expression relating torque and force for the rack and pinion transformer.

Ex. 5.2 Shown below is a (massless) fulcrum, which consists of a beam that pivots about a point. To the left of the point is a section of length A, and to the right is a section of length B. Derive the relations between the forces and velocities at the A and B ends of the fulcrum. Hint: the angular velocities on the A and B sides must be the same. Power is conserved.

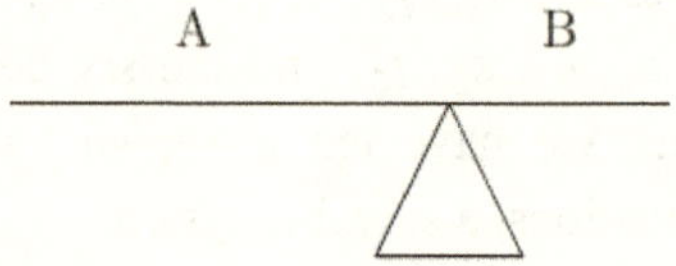

Ex. 5.3 Create a graph and equations for the system shown below. Gravity pulls downwards on the masses.

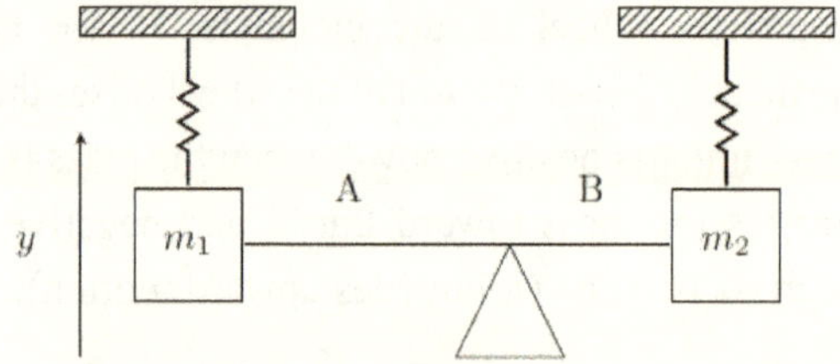

Ex. 5.4 In the previous exercise, place the springs below the masses so that support is offered from below. Draw the graph and derive the equations. Maintain the same orientations of your effort and flow variables.

Ex. 5.5 Draw the graph for the system shown below. The wheel with angular mass *I* can rotate but it cannot translate.

The top of the mass interacts with the wheel as a rack and pinion. The wheel has radius r.

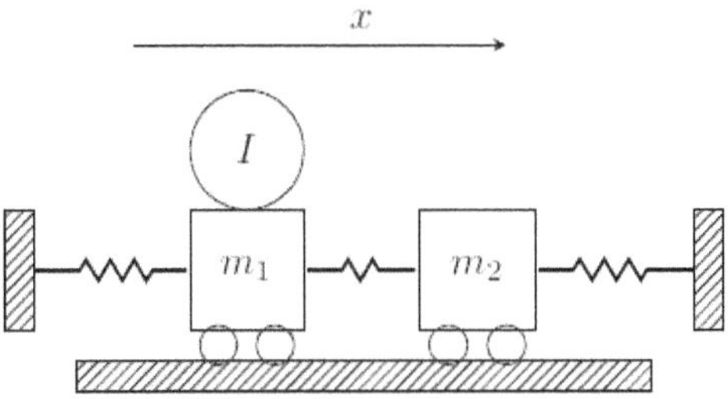

Ex. 5.6 Draw the graph for the system shown below. Hint: the point of connection between the damper and spring can move.

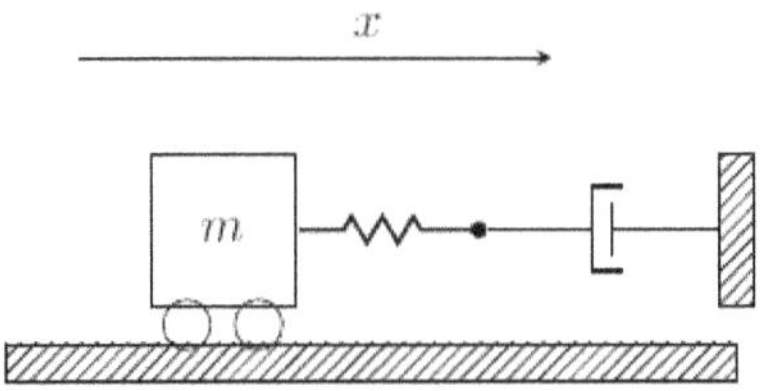

Ex. 5.7 In the previous exercise, remove all redundant edges and write the resulting set of equations. If $f_1 > 0$ is motion from left to right, how should we interpret $f_3 > 0$. Edge labels are shown in the exercise solutions.

Ex. 5.8 Using your graph from exercise 5.6, introduce a mass where the spring and damper are connected. Construct the simplified graph for this system as you did in exercise 5.7.

6. Electric circuits

From the simplicity of an incandescent light bulb to the complexity of a modern computer, electric circuits play a major role in almost every system built today. In this chapter, we discuss the basic principles of analog electric circuits that are built with resistors, inductors, capacitors, and transformers. This small set of components is sufficient to model a wide range of systems, including electric power networks, analog filters, and simple radios.

Flow in an electric circuit is the number of electrons moving past a point in a conducting material per unit time. The flow f has units of Coulombs (C) per second, where a Coulomb is equal to the charge of approximately 6.3×10^{18} electrons. The accumulated flow q is charge, and it has units of Coulombs. Fortunately, you don't need to remember this: one Coulumb per second is called an Ampere and this is the term used in practice to describe electric current.

Effort is electric potential measured in Joules per Coulumb (J/C). One Joule per Coulumb is called a Volt. Multiplying effort and flow we get

$$ef = \frac{\mathrm{C}}{\mathrm{s}}\frac{\mathrm{J}}{\mathrm{C}} = \frac{\mathrm{J}}{\mathrm{s}} = \mathrm{W}$$

The traditional notation in electrical engineering uses i for current and v for voltage. Consequently, in electrical engineering textbooks j is used for the imaginary number $\sqrt{-1}$. The traditional notation also conflicts with the use of v for velocity in mechanical systems and volume flow in hydraulic systems. In those places, v is a flow rather than an effort. Keep these differences in mind as you work through the examples.

0 and 1 junctions If you have never built an electric circuit, it may be helpful to think of conducting wires as pipes in a hydraulic network. The current flowing through the conductor is analogous to the volume of fluid flowing through the pipe. The pressure at the ends of the pipe is analogous to the electric potential at the ends of a wire.

A zero junction is a point where conducting wires meet. A zero junction cannot store charge. Therefore, the sum of currents flowing into the junction is equal to the sum of currents flowing out of the junction. There is a single electric potential at the junction.

A one junction is a section of conducting wire in the circuit. The wire has a single flow. If e_1 is the sum of electric potential entering the wire and e_2 the sum of electric potential leaving the wire, then conservation of power requires

$$f e_1 = f e_2$$

Hence, the sum of the efforts at the start of the wire must equal the sum of efforts at the end of the wire.

The zero junction is Kirchhoff's current law. In electrical engineering textbooks, this law is stated as 'the sum of currents entering a node in the electrical network must be zero'. The one junction is Kirchhoff's voltage law. This law is commonly stated as 'the sum of voltage drops in a closed loop must be zero'. While the current law is a simple restatement of the zero junction's property, the relationship of the voltage law to the one junction may be less clear. The examples will shed some light on this relationship.

Basic elements The Z element is an electrical resistor. In fact, every electrical element, including the wire that carries current between elements, resists flow through it. In practice, these parasitic resistances are so small in relation to resistances introduced intentionally that they can be neglected. The resistor is used to model any resistance that is considered important to the circuit's function.

For example, on a breadboard used for prototyping circuits, the resistances in the model appear as physically present electrical elements, called resistors, introduced specifically to impede the flow of current and produce a drop in voltage. You can see these resistors on the breadboard. Contrary to these intentional resistances, when modeling electrical power systems with long lines (imagine the transmission lines you might see crossing over a highway) it is necessary to include the resistance inherent in the conducting wires to obtain a useful result.

Electrical engineers have produced a standard system of symbols for drawing electric circuits. In this system, the resistor appears as

where R is the parameter of the Z element. Resistance is measured in Ohms, written as, for example, 2Ω.

The E element is a capacitor that stores power in an electric field. There are many types of capacitors, but in the standard symbols of a circuit drawing the capacitor appears as

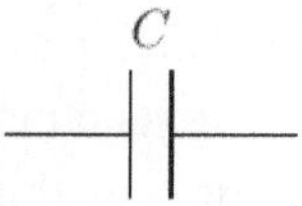

where C is the parameter of the E element. The capacitance C is measured in Farads, but because one Farad is an enormous capacitance, microFarads μF, and milliFarads mF are common in circuit designs.

The F element is an inductor that stores power in a magnetic field. As with the capacitor, there are many types of inductors, but in the standard symbols the inductor appears as

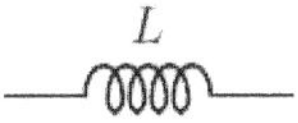

where L is the parameter of the F element. The inductance L is measured in Henries, with microHenries, μ H, and milliHenries, mH, being typical in circuit designs.

A source of effort appears where a voltage is applied to the circuit. A typical source of effort is a battery or an integrated circuit designed to supply constant voltage (a voltage regulator). A source of effort is drawn as

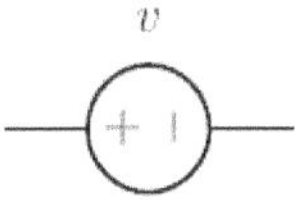

The voltage v is the difference in electric potential between the positive (high) side and the negative (low) side of the voltage source.

One or more nodes in a circuit are chosen to be ground. The ground has a voltage of zero. In many cases a ground is literally connected to the Earth, which provides a sink for current as it flows out of the circuit. A symbol commonly used for ground is

A source of flow appears where current is injected into the circuit. Solar cells are often modeled as a source of flow with an intensity depending on the amount of light falling on the cell's surface. Integrated circuits that supply constant current can be modeled as sources of flow. A common symbol for a source of flow is

The arrow shows the direction of positive flow.

Drawings to graphs The procedure for translating electric circuits into a system of equations has much in common with hydraulic systems. In many respects, the drawing conventions of the electrical engineer simplify our task. To begin, assign directions for the flow of current in each conducting wire of the network; that is, on each line or branch connecting a pair of nodes. In some cases, this will have been done for you. In others, you must provide the direction yourself.

Consistency will simplify the system of equations. Currents flow naturally[6] from high voltage to low voltage. Therefore, you will want to orient your currents away from the positive terminals of voltage sources and toward ground. The current

---

[6] This is true when the difference in voltage is sufficiently large. In semiconductor devices, there may also be a diffusion current caused by the uneven concentration of electrons in the device's material. This diffusion current seeks to redistribute the electrons such that their concentration is uniform, and it can flow against the 'natural' direction we describe here. This phenomenon is not relevant to any of the circuits that we will study.

along a branch containing a source of flow will always be oriented in the direction indicated by the current source.

With directions for the currents, we proceed to create a graph that reflects the shape of the circuit diagram. Each element in the diagram - resistor, capacitor, inductor, voltage source, and current source - becomes a one junction with the corresponding Z, E, F, SE, or SF element attached.

Each node in the circuit diagram becomes a zero junction between the one junctions of the elements. If we have a choice of direction for an edge going into or out of a junction (1 or 0), then the edge should point in the direction of flow that we have chosen. Several examples will illustrate the procedure.

Tank circuit The tank circuit consists of a voltage source, an inductor, and a capacitor connected in series. One place that you may encounter this circuit is in an amplitude modulated (AM) radio receiver. In an AM receiver, the capacitor is tuned such that the circuit resonates with the desired transmitting station and filters out nearby stations on the radio dial.

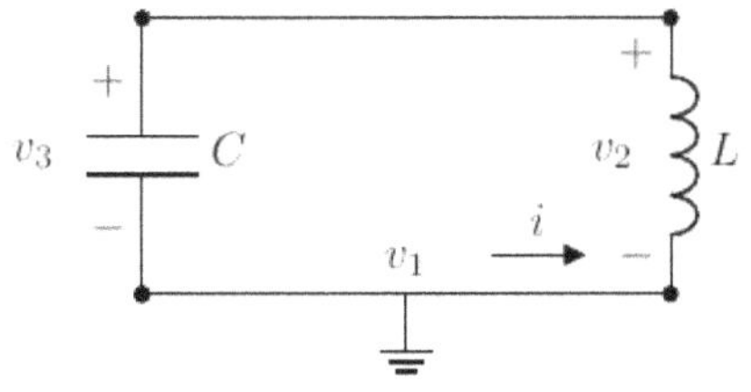

The direction of flow is indicated by the arrow with the label i . We have chosen the direction of flow to be counterclockwise, but the opposite choice would have been just as serviceable. To simplify our system of equations and adhere to standard practice, this direction for the current is used in each branch of the circuit. As we shall see, this choice is a natural consequence of the electrical laws embedded in our graph.

The voltage at ground is $v_1 = 0$. We have labeled the voltage across the inductor as v_2 and across the capacitor as v_3. The orientations of the voltages will be imposed by our graph. They are drawn here so that we may later see the connection between the graph's rules and voltage orientation.

This circuit has three elements: an E element with parameter C, an F element with parameter L, and an SE element that imposes a voltage of zero. We create our graph by placing a zero junction at each node in the circuit. To aid in our process, each node in the circuit is marked with a heavy dot. We place a one junction at each point where an element or line connects a pair of nodes, as such:

$$\begin{array}{ccccc} 0 & \longleftarrow & 1 & \longleftarrow & 0 \\ \downarrow & & & & \uparrow \\ 1 & & & & 1 \\ \downarrow & & & & \uparrow \\ 0 & \longrightarrow & 1 & \longrightarrow & 0 \end{array}$$

Next, we attach an element to each 1 junction corresponding to the place in the circuit diagram where it appears. If no element is along a line then we leave the one junction by itself. This step completes the circuit graph. We have labeled the edges that will persist as we simplify the graph.

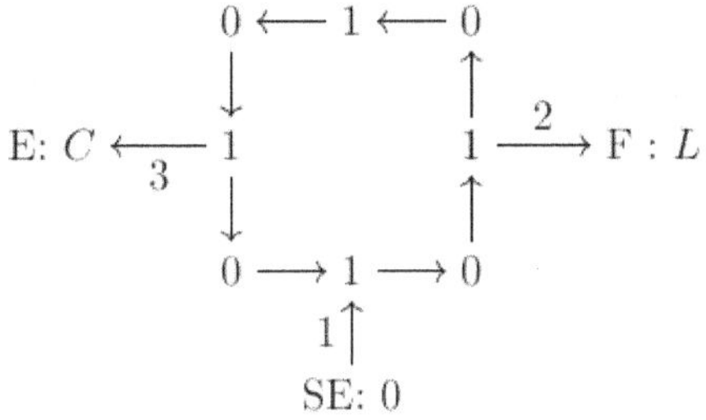

To simplify the graph, start at edge 1 with the flow f_1. Moving counterclockwise, the first zero junction that we encounter requires $f_1 = f_2$. Continuing counterclockwise, we find $f_2 = f_3$ and so there is a single flow $f = f_1 = f_2 = f_3$. The simplified graph is thus

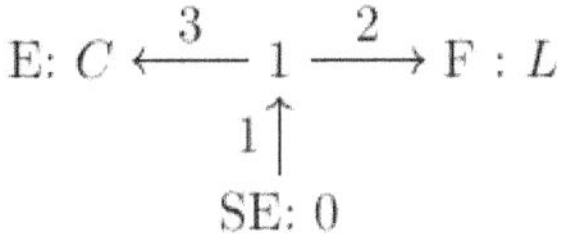

The system of equations produced by this graph is

$$\begin{aligned} \dot{e}_3 &= f/C \\ \dot{f} &= e_2/L \\ e_1 &= 0 = e_2 + e_3 \end{aligned}$$

From this system of equations we can deduce that $e_2 = -e_3$ and so, like the current, there is a single voltage in this circuit. The equations are

$$\begin{aligned} \dot{e} &= f/C \\ \dot{f} &= -e/L \end{aligned}$$

Mapping efforts and flows back to our drawing we have $i = f$ and $v_2 = v_3 = e$.

To see how the voltages are imposed, first consider the case where $f > 0$ and $e > 0$. The power flowing into the capacitor is fe, which is positive. To produce this positive power, the current must be flowing from the high voltage side of the capacitor to the low voltage side; that is, in the direction of the drop in voltage. Hence, the orientation that we placed in our drawing.

If fe is the power that flows into the capacitor, then conservation of power requires that $-fe$ be the power that flows into the inductor. The inductor is discharging power through its positive terminal, in opposition to the voltage drop, but in the direction of flow. For this reason, the single effort in our model requires the inductor voltage to have the same orientation as the capacitor voltage so that one provides power while the other absorbs power.

Keeping $e > 0$ and reversing the direction of the current so that $f < 0$, we see the capacitor and inductor change roles. Now, the capacitor provides power and the inductor absorbs power. This hints at the name 'tank' for this circuit. Power sloshes back and forth, like water in a tank, without ever spilling over the sides.[7]

Transmission line The circuit below models an electrical transmission line that connects a generator, which supplies current, to an electrical machine that we have modeled with a resistor (for example, an incandescent light build can be modeled as an electrical resistance). The transmission line consists of a resistor in series with an inductor. The resistor models electrical resistance in the conducting medium of the wire. The wound wiring of a long transmission line acts as

---

[7] Of course, this model is an idealization. The wires between components and the components themselves offer some small resistance to flow. In time, this resistance will dissipate the power as heat.

an inductor, storing energy in the magnetic field produced by the moving electrons as they accelerate while following the curve of the winding (this process is outside the scope of our model!).

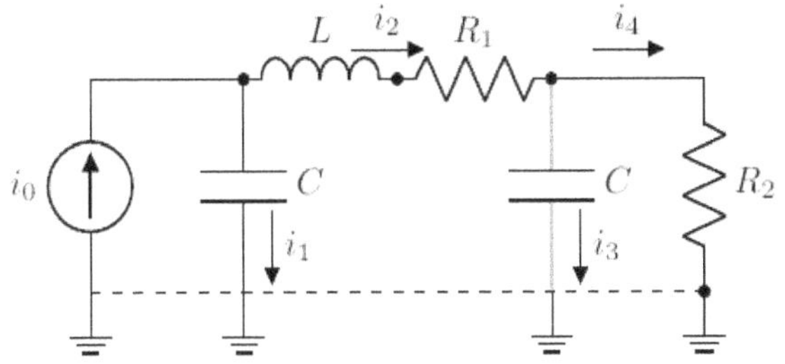

At each end of the transmission line is a capacitor. The electrical charge on the line is separated from the earth by the air. This forms a natural capacitor that stores energy in the electrical field created between the conductive line and earth. This process is outside the scope of our model but its effect on currents and voltages in the system is adequately represented by a pair of capacitors.

We will abbreviate our modeling procedure by selecting just those points in the circuit where elements are joined. Doing so eliminates redundant zero and one junctions that our rote procedure would have introduced. The points of connection that appear as zero junctions in our graph are indicated with black dots.

A single source of effort will model the four grounds in our circuit. Each ground imposes a voltage of zero where it is connected. We can imagine the grounds being connected by a wire and select one ground at which to place the black dot. The imagined, dashed line and dot are indicated in the figure.

We say imagined because, though the grounds are electrically connected (literally to the ground or, in portable electronics, to a metal plate that serves the same purpose), this common point of connection is often omitted from engineering diagrams. By making it explicit, the justification for a single source of effort in our graph is apparent.

Each node in the circuit diagram becomes a zero junction in our graph. Between zero junctions we place a one junction where circuit elements appear. We attach to each one junction the appropriate circuit element.

We choose our edge directions to match the flows shown in the circuit diagram. The flows move away from the source of flow, our generator, and towards the source of effort, our ground. Along each path in the graph, flow enters the one junction from the zero junction that is upstream and leaves the one junction towards the zero junction that is downstream. The resulting graph is as follows:

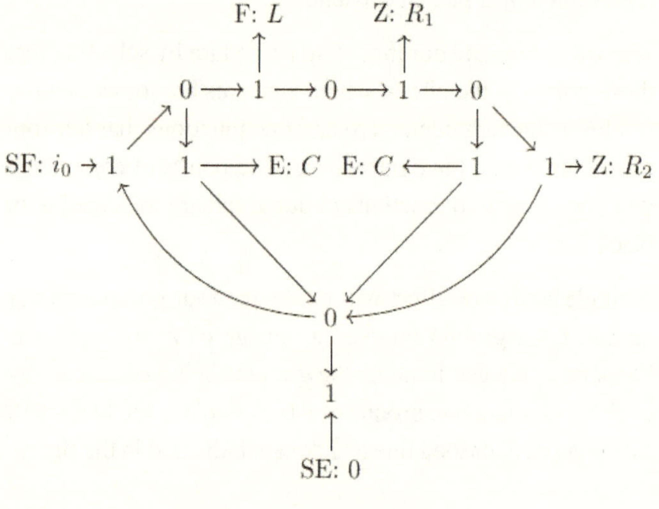

The graph closely resembles the circuit. The zero junction corresponding to each dark dot in the circuit diagram provides the voltage at that point relative to the ground. Each one junction provides the current flowing through the attached element.

Because the ground voltage is zero, we can eliminate this element from the graph without altering its system of equations. After doing so and then eliminating junctions with degree two, we arrive at the simplified graph shown below. Labels have been added to the edges of this simplified graph so that we may create its system of equations.

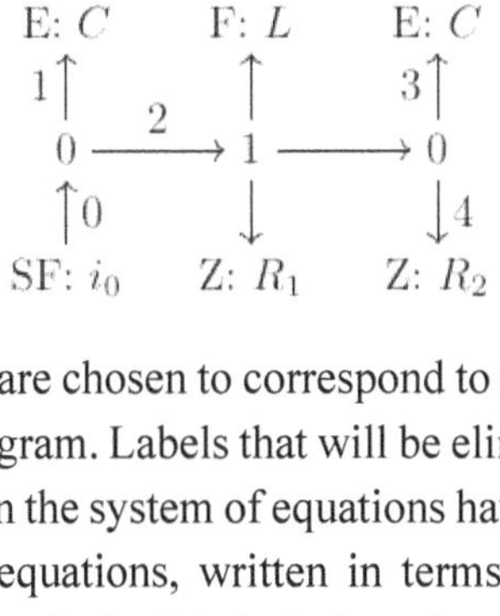

The edge labels are chosen to correspond to the flows shown in the circuit diagram. Labels that will be eliminated because of equivalency in the system of equations have been omitted. The system of equations, written in terms for efforts and flows and using only the labeled edges, is

$$\dot{e}_1 = (f_0 - f_2)/C$$
$$\dot{f}_2 = (e_1 - e_3 - R_1 f_2)/L$$
$$f_4 = e_3/R_2$$
$$\dot{e}_3 = (f_2 - f_4)/C$$

Circuit with a transformer An electrical transformer couples two circuits through a magnetic field. The magnetic field is produced by current moving through coils of wire wound on two sides of a ferrous ring. One set of coils is the transformer's primary side and the other is its secondary side.

If the current through the transformer changes in time, then this device exhibits a constant of transformation equal to the ratio of windings on the primary and secondary sides.

The physical process is called mutual inductance, and it relies on Faraday's law, which relates the rate of change in magnetic flux and voltage. The rate of change in magnetic flux is proportional to the rate of change of the current through the coil, and so we require a changing current. Transformers of this type are used in alternating current circuits where i and v are sinusoidal.[8]

Our example will use a source voltage

$$v(t) = \sin(\omega t)$$

This time varying voltage induces a time varying current through the transformer. Thereby, we satisfy the conditions required by Faraday's law to permit the transformer to act in accordance with our model.

Figure 11 shows a schematic of the transformer. The primary side has n_1 turns in its windings and the secondary side has n_2. The voltage v_1 is related to the voltage v_2 by

$$n_1 v_2 = n_2 v_1$$

The constant of transformation for the transformer element in a graph is

$$T e_2 = e_1$$

and so the constant of transformation is

---

[8] The simplicity of alternating current (AC) transformers is one of the reasons that Tesla's AC electricity beat out Edison's direct current electricity as the foundation for electric power systems today.

$$T = n_1/n_2$$

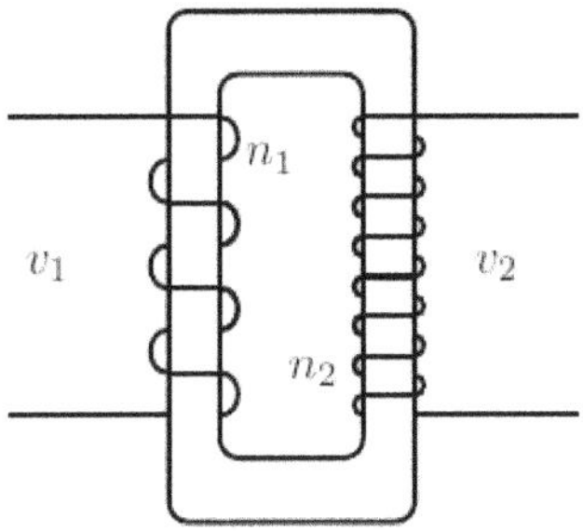

Figure 11. Schematic of a transformer.

The figure below shows an electric circuit with a transformer. The figure is followed by the graph for this circuit. The edges adjacent to the transformer go into the primary side and out of the secondary side. The voltage on the primary side is v_1 and on the secondary side is v_2. As in the circuit, the transformer element in the graph connects the primary coils and the secondary coils. The points of connection are modeled with one junctions.

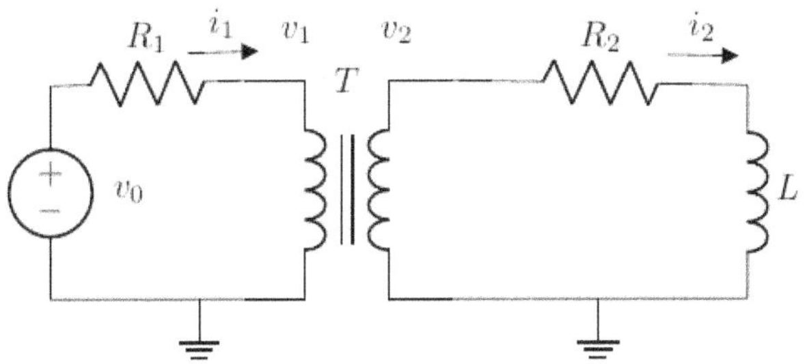

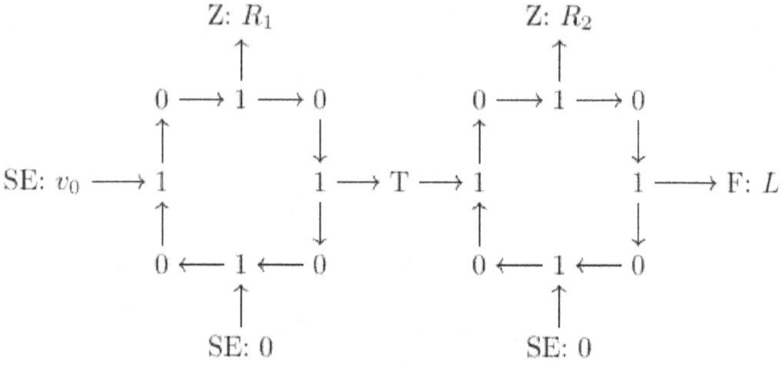

We can simplify the graph by removing junctions of degree two (except the transformer!) and combining one junctions. Doing this leaves us with a pair of one junctions connected by the transformer. The graph is labeled to be consistent with the voltage and current labels in the circuit diagram. Edge 0 supplies voltage v_0, edge 1 has voltage v_1, and edge 2 has voltage v_2. There are two currents, as indicated by the two one junctions in the graph. The one junction on the left has current i_1 and the other on the right i_2.

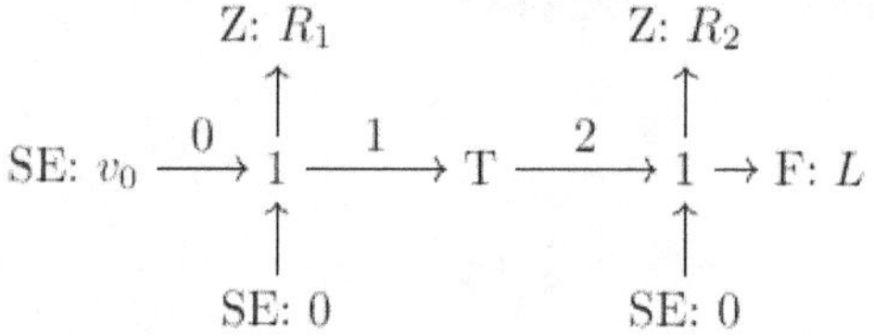

The system of equations from this graph, written in terms of efforts and flows rather than voltages and currents, and simplified to remove trivial equalities and the sources of effort that supply zero voltage, is

$$
\begin{aligned}
&e_1 = e_0 - f_1 R_1 = \sin(\omega t) - f_1 R_1 \\
&\dot{f}_2 = (e_2 - f_2 R_2)/L \\
&e_1 = T e_2 \\
&T f_1 = f_2
\end{aligned}
$$

Exercises

Ex. 6.1 Derive units for the R parameter of the resistor, the C parameter of the capacitor, and the L parameter of the inductor in terms of current (Amperes) and Volts.

Ex. 6.2 Create a graph for the circuit shown below.

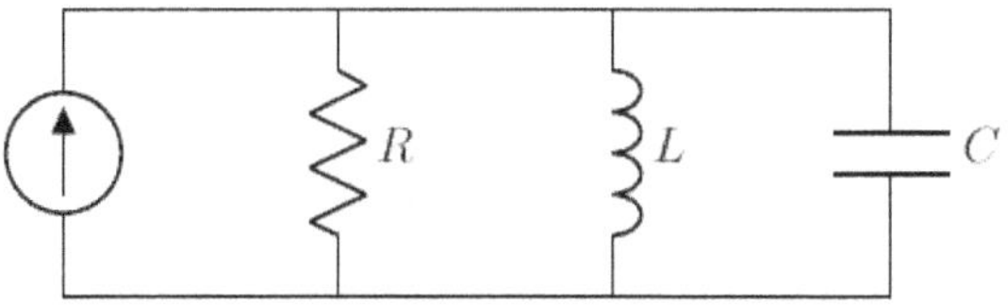

Ex. 6.3 Create a graph for the following circuit.

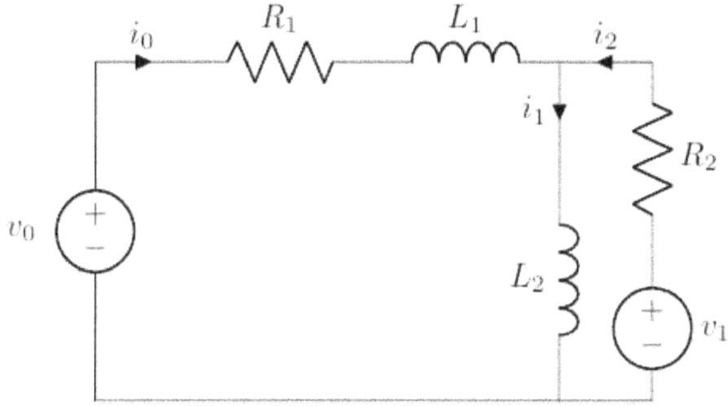

Ex. 6.4 Change the direction of the current i_2 in the previous exercise. How should your graph change?

Ex. 6.5 Show that two branches in a circuit that share common points of connection at each end must have equal changes in voltage; these are called parallel branches. For example, the branches carrying i_0 and i_1 in exercise 6.3 are in parallel, as are the branches carrying i_1 and i_2. Hint: refer to the diamond structure in Chapter 2.

Ex. 6.6 In the bond graph for the circuit with transformer that was shown in this chapter, change the orientation of the edges pointing into the transformer. What is the constant of transformation in this case? Show that, with your choice, you obtain the same system of equations for the circuit.

Ex. 6.7 Draw an electric circuit that has a graph matching the example of the constant power pump in the chapter on hydraulics. Create any symbol you like for the pump.

Ex. 6.8 Draw a graph of the circuit below.

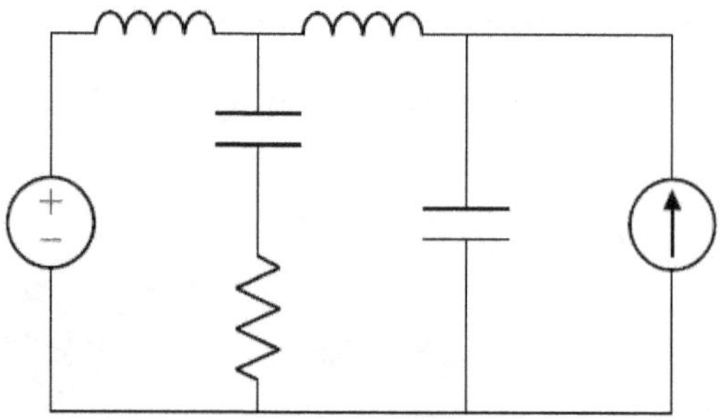

Ex. 6.9 Draw a graph of the circuit below.

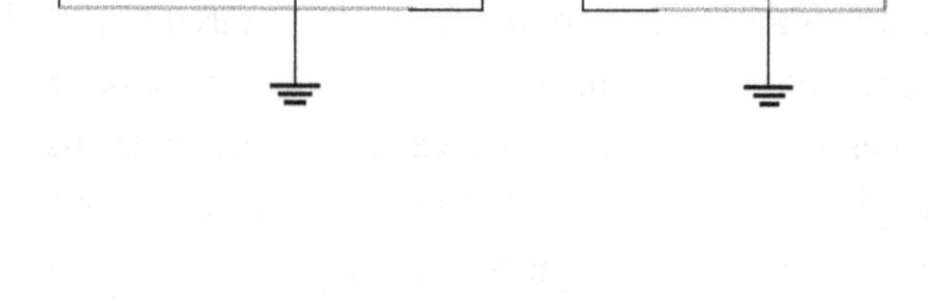

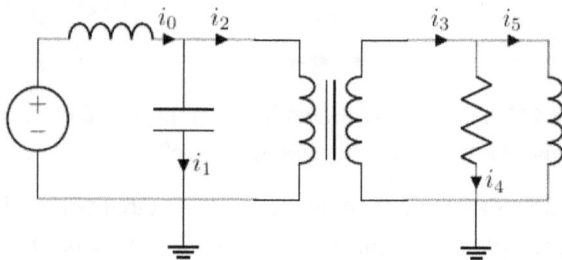

7. Transformers and gyrators

We have seen several instances of transformers within a single domain. For instance, the interlocked gears of a rotating mechanical system and the electrical transformer. Those devices have on each edge the same physical interpretation of effort and flow. A transformer describing two interlocked gears relates the torque and angular velocity of the driving gear to the torque and angular velocity of the driven gear. The electrical transform relates voltage and current on the primary side to voltage and current on the secondary side.

To this point, only the transformer that models a rack and pinion relates unlike variables. The torque and angular velocity of the gear is related to the force and linear velocity of the rod. In this chapter, we expand our list of transformers that relate like variables and examine several more transformers and gyrators that relate unlike variables. The latter are particularly useful as they permit us to link systems from separate domains.

Electric motor The gyrator is a simple model of an electric motor. A concrete example of an electric motor that can be modeled as a gyrator is the type of permanent magnet motor shown in Figure 12, which is available from almost any electronic hobby shop. A motor can transform mechanical power into electrical power or can be used as a mechanical drive that operates from electrical power.

The motor in Figure 12 has two electrical pins connected to wire coils inside the housing. The shaft can be attached to a wheel, turbine, pulley, or other rotating machine. When a voltage appears across the electrical pins, this causes a current to flow through the coils in the motor and induces a torque on the shaft. Similarly, if we apply a torque to the shaft, this induces a current to flow through the motor and causes a voltage to appear across the pins.

A gyrator models the relationship between voltage v, torque τ, angular velocity ω, and current i. To use the motor in a model, we must decide which are inputs to the gyrator, attached to the edge pointing into the element, and which are outputs. The convenient choice will be dictated by the system being modeled.

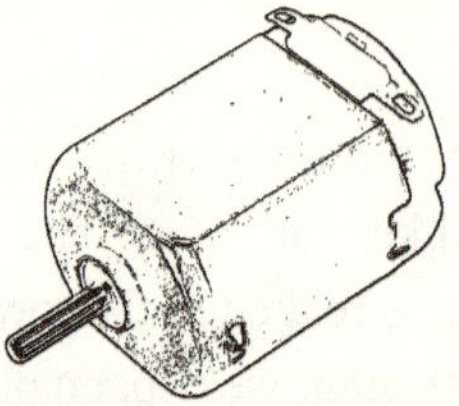

Figure 12. A permanent magnet motor.

The constant of gyration can be inferred from the information provided with the motor. This particular motor is designed to run at 440 radians per second with 1.5 volts across the pins. Let us assume the motor is being used as a mechanical drive so that electrical quantities are input and mechanical quantities are output. In this configuration,

$$\tau = Gi$$
$$v = G\omega$$

We can calculate G as

$$G = \frac{v}{\omega} = \frac{1.5 \text{ V}}{440 \text{ rad/s}} = 0.00341$$

An electric train with speed control will illustrate the use of an electric motor in a model. The train follows a straight, but hilly, stretch of track. The position of the train along the track is x and the grade of the track - its steepness - in radians is a function $\phi(x)$ that is given to us. We might obtain ϕ from a topographic map showing how the track follows the terrain.

A positive angle indicates rising ground. The train with mass m experiences a force due to gravity

$$F_g(x) = -mg\sin(\phi(x))$$

This force is parallel to the track and accelerates (on a downward slope) or decelerates (on an upward slope) the body of the train. The force of gravity when the train climbs is in opposition to the motor force. When the train descends, it is in the direction of the motor force.

The train also encounters resistance b to motion due to friction with the track, air drag, and other inefficiencies. Force supplied by the motor to the train's wheels pushes the train forward or causes it to brake. The wheel of the train has a radius r. Its effect to drive the train along the track can be modeled as a rack and pinion transformer.

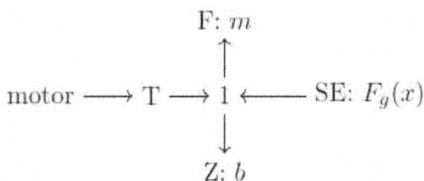

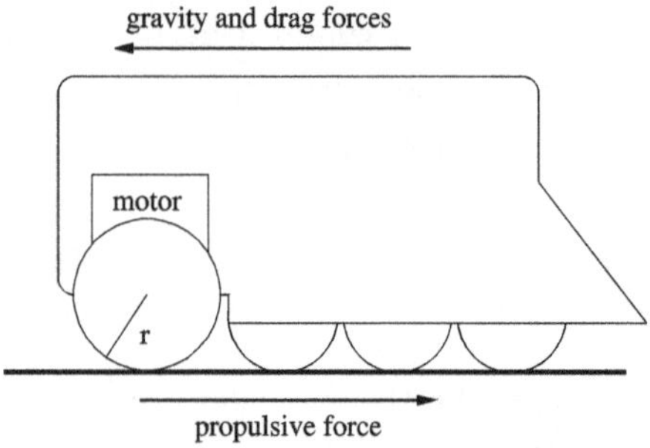

Figure 13. Elements of the electric train.

These mechanical elements are illustrated in Figure 13. The graph fragment for the mechanical elements in its simplified form is shown below. We have included a place holder 'motor' for the electric motor. The edges adjacent to the motor and transformer are oriented so that the power flows from the motor to the train.

$$
\begin{array}{ccccc}
 & & \text{F}: m & & \\
 & & \uparrow & & \\
\text{motor} \longrightarrow & \text{T} \longrightarrow & 1 & \longleftarrow & \text{SE}: F_g(x) \\
 & & \downarrow & & \\
 & & \text{Z}: b & &
\end{array}
$$

An electric motor under active control supplies propulsive force to the wheels of the train. The wire coils inside the motor produce a magnetic field that interacts with the magnetic field of the permanent magnet to produce torque. The wire coils can be modeled as an inductance L that is in series with the voltage source.

A computer monitors the speed v of the train and adjusts the voltage V applied across the motor pins to maintain a speed s. In this model, we assume some control function h in the form

$$V = h(v, s)$$

but leave the design of h to a control engineer. The control function appears as a source of effort in our model.

Appending the graph for the motor and controller to the train and labeling the edges that will appear in our equations, we arrive at

$$
\begin{array}{ccccccccc}
 & & \text{F: } L & & & & \text{F: } m & & \\
 & & \uparrow & & & & {}^{1}\uparrow & & \\
\text{SE: } h(v,s) & \longrightarrow & 1 & \xrightarrow{4} & \text{G} \xrightarrow{3} \text{T} & \xrightarrow{2} & 1 & \longleftarrow & \text{SE: } F_g(x) \\
 & & & & & & \downarrow & & \\
 & & & & & & \text{Z: } b & &
\end{array}
$$

The flow f_4 is the current through the motor and the effort e_4 is the voltage that relates to the speed of the motor shaft. The motor acts as a gyrator as it converts the current and voltage into rotational motion of the motor shaft. The flow f_3 is the angular speed of the motor shaft and the effort e_3 is its torque.

The train's wheel is attached to the motor shaft. The wheel and the track act as a rack and pinion transformer to move the train forward. The effort e_2 is the liner force propelling the train along the track. The flows $f_2 = f_1$ are the speed of the train along the track. The accumulated flow q_1 is the position of the train along the track. The system of equations written in terms of efforts and flows is

$$
\begin{aligned}
\dot{f}_4 &= (h(f_1, s) - e_4)/L && \text{motor current} \\
\dot{f}_1 &= (e_2 + F_g(q_1) - f_1 b)/m && \text{train speed} \\
\dot{q}_1 &= f_1 && \text{train position} \\
e_2 &= r e_3 && \text{driving force \& torque} \\
e_3 &= G f_4 && \text{motor torque \& current}
\end{aligned}
$$

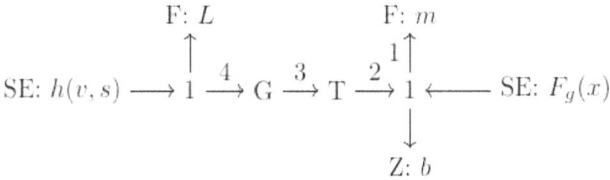

Hydraulic lift Figure 14 illustrates a hydraulic lift. This system uses a small mechanical force to lift a large load. The lifting principle is modeled as a transformer.

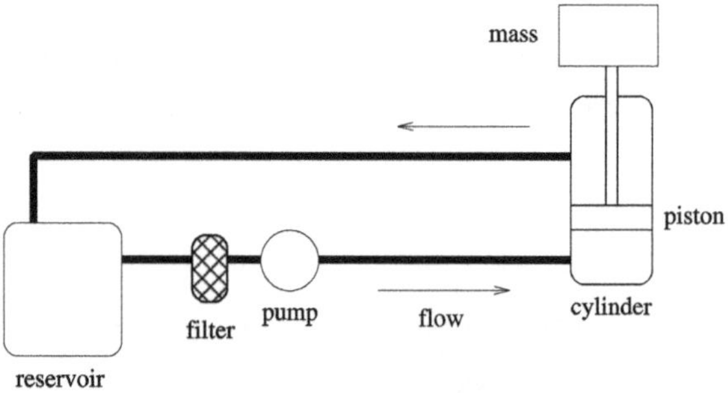

Figure 14. Schematic of a hydraulic lift configured to lift a mass.

A piston separates the lifting chamber into an upper half and a lower half. When the circuit is configured to lift a mass, the pump pushes fluid from the reservoir into the lower chamber. A line from the upper chamber allows the fluid that is forced out of the upper chamber to flow back into the reservoir, thereby closing the hydraulic circuit.

For the purposes of our model, we may take variations of pressure in the reservoir (for example, caused by changes in the fluid level) to be small in relation to the pressures acting elsewhere in the system. That is, the reservoir acts as a source of constant pressure. It is convenient to discuss all other pressures relative to the reservoir pressure. Doing so allows us to set its value to zero.

If the lines contain relatively little fluid when compared to the other elements in the system, then the inertia of the fluid in these lines can be ignored as we build the model. The chambers above and below the piston accumulate fluid to create pressure. These are modeled as E elements.

The filter is modeled as a nonlinear resistor in the form

$$Z(q) = Z_0 e^{kq}$$

The quantity q is the accumulated flow through the resistor. The parameters k and Z_0 are constants that depend on the filter design and cleanliness of the fluid.

The force to lift the piston is produced by the difference in pressure between the upper and lower sides of the piston plate that is submerged in the fluid of the cylinder. Let P_1 be the pressure in the upper chamber and P_2 the pressure in the lower chamber. Let A be the surface area of the piston plate. Then the hydraulic force F_1 pushing down on the plate is

$$F_1 = P_1 A$$

and the hydraulic force F_2 pushing up on the plate is

$$F_2 = P_2 A$$

This is a pair of transformers. The transformation constant is A if the transformer is oriented such that the incoming edge is force and the outgoing edge is pressure. If the edges are reversed, then the constant of transformation is $1/A$. The forces F_1 and F_2 act to lift the total mass m of the piston and its load. The net lifting force is opposed by the force of gravity $-mg$.

The graph for the hydraulic aspect of the lift is shown below with a placeholder for the piston. Several junctions of degree two are retained so that the diagram of the lift and the graph are visually similar and the correspondence between diagram and graph is clear. The direction of flow is from the reservoir (SE), through the filter (Z) and pump (SF), and into the lower chamber (E_2). As the piston rises, fluid flows out of the upper chamber (E_3) and into the reservoir.

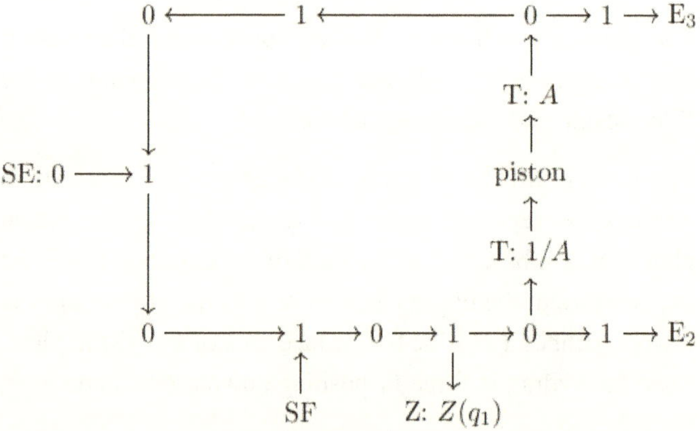

The power of the fluid flowing into (or out of) a chamber is split into two parts. The first part increases the pressure in the chamber. The second part works to move the piston. This split is modeled with a zero junction, which captures the single pressure in the chamber and the division of work by a division of flow. It is in cases like these that the diagram does not so clearly direct the model construction and the clarity and simplicity of the electrical engineer's symbology can be best appreciated.

Moving around the hydraulic part of the graph, we find several junctions of degree two. We simplify the graph by removing these and combining the adjacent one junctions. A further simplification removes the source of effort, which only introduces a zero into our equations. These simplifications produce the graph shown below.

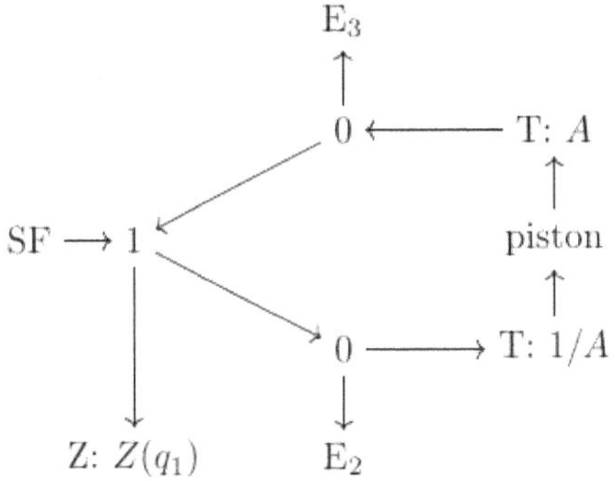

The piston consists of its mass and the mass that is being lifted, the force of gravity acting on the total mass, and the forces supplied by pressure in the upper and lower chambers. The graph fragment for this mechanical aspect of the lift is shown below.

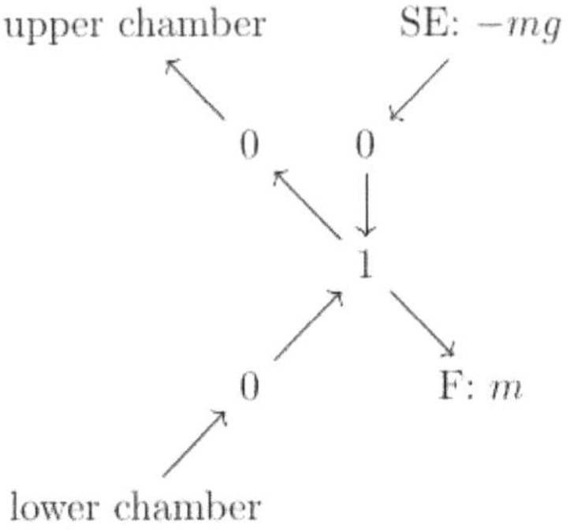

The direction of flow is chosen to be consistent with our choice in the hydraulic subsystem. Positive velocity is upward. Positive pressure in the lower chamber causes a positive upward motion and delivers positive power to the piston. As the piston rises, power is delivered to the upper chamber, causing pressure to increase and fluid to flow out of the chamber. Gravity supplies positive power when the piston moves downward.

Introducing the graph fragment of the piston into its placeholder within the hydraulic graph and simplifying once more produces the graph for the complete lift system.

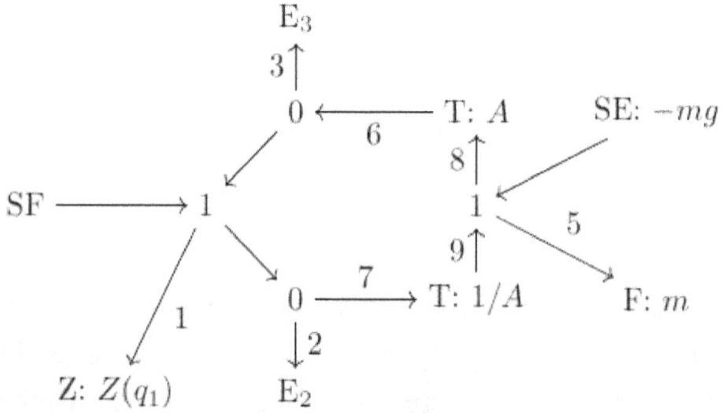

The flow f_1 is provided by the pump. The pressure equations for the upper and lower chambers and for the motion of the piston are

$$\begin{aligned}
\dot{e}_3 &= (f_6 - f_1)/E_3 \\
\dot{e}_2 &= (f_1 - f_7)/E_2 \\
\dot{f}_5 &= (e_9 - e_8 - mg)/m \\
Af_5 &= f_6 = f_7 \\
e_9/A &= e_2 \\
Ae_3 &= e_8
\end{aligned}$$

By modeling the pump as a source of flow, we have assumed that the accumulation of silt in the filter does not impede the operation of the lift in any significant way. However, the filter is not without effect. The power extracted from the flow by the resistance of the filter must be supplied by the pump. This power is

$$\begin{aligned}
f_1 e_1 &= f_1 Z(q_1) \\
\dot{q}_1 &= f_1
\end{aligned}$$

Eventually, this power plus the power needed to lift the mass will exceed the pump's rated power and the hydraulic lift will cease to operate in accordance with our model.

Attitude control A spinning gyroscope has an angular moment of inertia $\bar{I}$. The angular speed of the gyroscope is a vector $\bar{\omega}$. We will assume that the gyroscope is rotating at a constant speed and very close to its z-axis so that the angle α between $\bar{I}$ and $\bar{\omega}$ is small enough that $\sin(\alpha) \approx \alpha$.

When the gyroscope lies almost entirely along the z-axis, the vector quantities have two zero elements and a single nonzero element. Let the nonzero element of $\bar{I}$ be I. Likewise, let the nonzero element of $\bar{\omega}$ be ω. With these and the small angle assumption, the angular momentum p of the gyroscope is approximately a scalar quantity

$$p = \bar{I} \times \bar{\omega} \approx I\omega\alpha$$

Taking the time derivative we get

$$\dot{p} = I\omega\dot{\alpha}$$

The term $\dot{\alpha}$ is the angular acceleration of the gyroscope as it tilts in the direction perpendicular to its motion. This is a flow. Recalling the derivation of mechanical momentum, the term $\dot{p}$ is an effort. Specifically, it is the torque that induces the angular acceleration $\dot{\alpha}$. Define $G = I\omega$ and we have obtained a gyrator

$$e = Gf$$

Now consider a ship at sea. Its cross section, when viewed from the stern, has a width, or beam, b. Against each side of the ship, at a distance of half the beam from the center line, the movement of the sea imparts a force. This force causes the angular mass of the ship to roll about its center.

The graph of this system has two sources of effort: the forces on the port and starboard sides of the ship that are caused by the ocean's motion. The beam length of the ship acts as a fulcrum to transform these ocean forces into a roll about the ship's center of mass. The simplified graph of this system is shown below. The constant of transformation $T = b/2$.

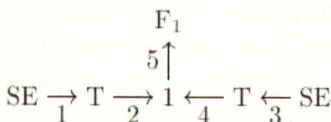

$$\begin{array}{ccccccccc} & & & & F_1 & & & & \\ & & & & 5\uparrow & & & & \\ SE & \xrightarrow[1]{} & T & \xrightarrow[2]{} & 1 & \xleftarrow[4]{} & T & \xleftarrow[3]{} & SE \end{array}$$

The roll of the ship is evident in the equation of motion

$$\dot{f}_5 = (2e_1/b + 2e_3/b)/F_1$$

The forces e_1 and e_3 are the action of the sea upon the ship.

Now suppose that we mount a gyroscope in the hull of the ship. The gyroscope wheel is mounted so that its angle α is free to rotate fore and aft. The rotation ω of the gyroscope wheel is about the mast of the ship. We can model the mounting of the gyroscope to the hull with our gyrator element as shown below.

$$\begin{array}{ccccccccc} & & & & F_1 & & & & \\ & & & & 5\uparrow & & & & \\ SE & \xrightarrow{1} & T & \xrightarrow{2} & 1 & \xleftarrow[4]{} & T & \xleftarrow[3]{} & SE \\ & & & & \downarrow 6 & & & & \\ & & & & G & & & & \\ & & & & \downarrow 7 & & & & \\ & & & & F_2 & & & & \end{array}$$

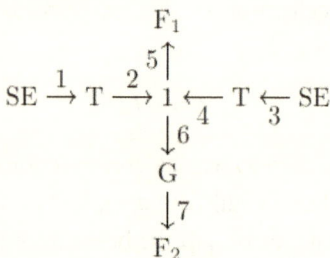

Here, F_2 is the angular mass of the gyroscope and $f_7 = \dot{\alpha}$. Now the forces of the ocean are counteracted by the torque from the gyroscope, as is evident in the new equations of motion.

$$\dot{f}_5 = (2e_1/b + 2e_2/b - Gf_7)/F_1$$
$$\dot{f}_7 = (Gf_5)/F_2$$

Exercises

Ex. 7.1 Derive the constitutive equations for the crank shaft and wheel shown below. The points of connection highlighted with black dots are free to rotate. The wheel has radius r and the connecting bar has length l. Hint: use the law of cosines to describe the triangle formed by the three heavy dots.

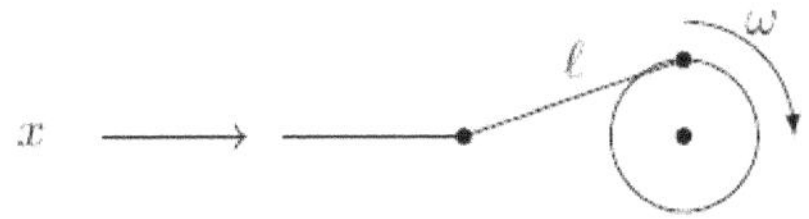

Ex. 7.2 A reciprocating pump is illustrated below. This pump moves fluid out of the reservoir and through the delivery pipe. You may assume that the reservoir and the outlet are at the same constant pressure. When the crank is turned to pull the piston right, the pressure in the piston chamber is reduced. This causes the lower valve to open, the upper valve to close, and fluid accumulates in the chamber. When the piston is pulled left, the pressure increases with the opposite effect. Draw a graph for this system. Model the valves as diodes (see the chapter on electric circuits). The crank wheel is turned at a constant speed. The wheel has radius r and the piston has surface area A.

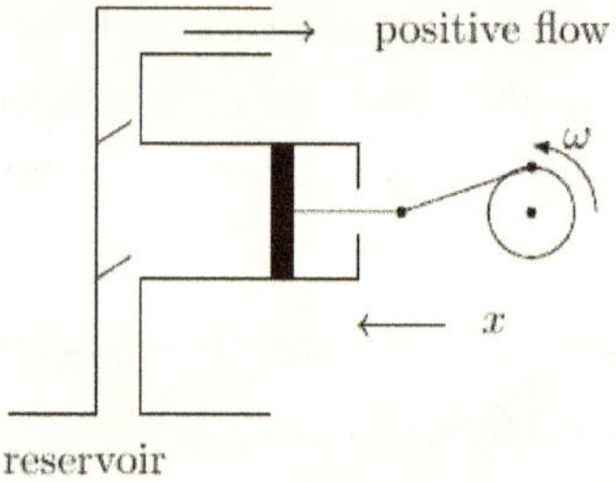

Ex. 7.3 Explain why a source of constant power is infeasible for this model of the reciprocating pump.

Ex. 7.4 Redraw the graph of the hydraulic lift by replacing the zero junctions adjacent to the transformers with one junctions. Attach E_1 and E_2 directly to their respective one junction. Show that this error can be detected by deducing that the pressures in the upper and lower chambers increase when flow is positive; that is, both chambers fill rather than the lower chamber filling and upper chamber draining.

Ex. 7.5 Show that the graph fragment

$$\xrightarrow{1} \mathrm{G} \longrightarrow \mathrm{T} \xrightarrow{2}$$

imposes the relationship

$$e_1 = (G/T)f_2$$
$$f_1 = (T/G)e_2$$

Ex. 7.6 Derive the constitutive equation for the idealized block and tackle shown below. You may assume all components are massless and perfectly stiff (no springiness in the cable). The left wheel (called a block) is firmly anchored to the ceiling. The right block is free to move up and down. The load (lift force) is firmly attached to this moving block. The dashed line is the cable.

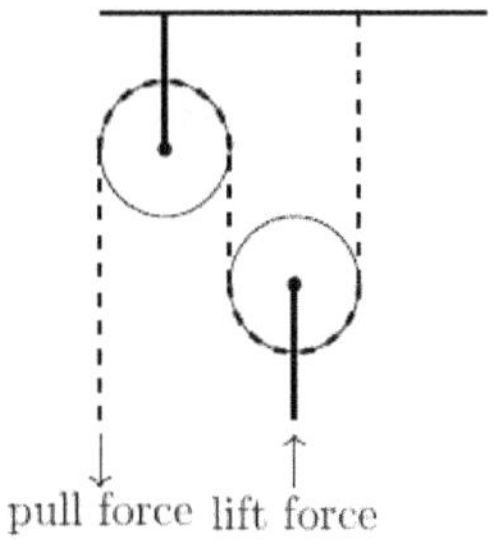

Ex. 7.7 Attach a mass to the lifting end of the block and tackle in the previous exercise. An electric motor like that used in the locomotive example supplies the pulling force. Construct the graph for this system.

8. Thermal systems

A rudimentary, but still useful, form of heat transfer can be expressed using our graph method. To arrive at quantities for effort and flow, consider a small volume of space. We assume that changes in pressure within the volume are negligible and that no mass enters or leaves the volume.

The temperature within the volume changes when there is a flow of heat through its surface. A natural choice for a flow variable is the flow of heat, analogous to current and volume flow. Temperature is a natural choice for effort, analogous to voltage and pressure. These are the primary variables in a standard study of heat transfer.

The flow of heat P through the surface of the volume has units of power. The temperature T is measured in degrees Kelvin (K). Clearly, the product TP does not have units of power. If we wish to incorporate thermal effects into a graph model containing electric, hydraulic, and mechanical systems, then some alternative is needed.

We retain T as our effort variable and introduce S, called entropy, as our accumulated flow. The flow $dS/dt = \dot{S}$ is called entropy flow. For $T\dot{S}$ to have units of power, entropy must have units

$$S = \frac{\mathrm{J}}{\mathrm{K}}$$

and entropy flow have units

$$\frac{dS}{dt} = \frac{\mathrm{J}}{\mathrm{K} \cdot \mathrm{s}}$$

Unlike accumulated flow in our other domains, the accumulated entropy in a closed system always increases. Consider a flow of power between two volumes, a warm volume with temperature T_h and a cool volume with temperature T_c. Heat cannot enter or leave our system and is only exchanged between these volumes: the system is closed.

Heat flows from hotter to colder temperatures. Conservation of power requires that the flow of heat into the cooler volume be equal to the flow of heat out of the warmer volume. Assigning entropy flow $\dot{S}_c$ to the cooler volume and $\dot{S}_h$ to the warmer volume, it follows from conservation of power that

$$T_c\dot{S}_c + T_h\dot{S}_h = 0$$

Temperature on the Kelvin scale is strictly positive.[9] Accordingly, $\dot{S}_c > 0$ because power flows into the cooler volume, $\dot{S}_h < 0$ as power flows out of the warmer volume, $|\dot{S}_c| > |\dot{S}_h|$ is necessary for the equality to hold. The total rate of change in entropy is

$$\dot{S}_c + \dot{S}_h > 0$$

This irreversible change in entropy is in stark contrast to an electric current that might flow between the two volumes, and it prevents a perfect analogy. The charge lost by one volume must match the charge gained by the other. The total

---

[9] The Kelvin scale measures absolute temperature. At zero Kelvin all microscopic motion has stopped, and it is impossible to achieve a lower temperature. This absolute temperature is unlike temperature in degrees Celsius, which is measured relative to the freezing point of ice, or degrees Fahrenheit, which is measured relative to the freezing point of a mixture of water, ice, and salt.

change in charge is zero. However, the total entropy in our closed system always increases when there is a flow of heat.

This has important consequences when the source of heat is a Z element in one of our other domains. Consider the power

$$P = f^2 Z$$

that is dissipated by a Z element. This heat is deposited into the Z element at a temperature T with entropy flow $\dot{S}$ so that

$$f^2 Z = T\dot{S}$$

Clearly, $\dot{S} > 0$ whenever $f \neq 0$. Entropy always increases during this exchange of energy.

Moreover, we cannot recover the energy transformed into heat by the Z element. Consider the small quantity of heat energy $Pdt = dQ$ deposited in the Z element in infinitesimal time dt. The corresponding change dS in entropy is

$$dQ = TdS$$

We have already established that $dS > 0$. Therefore, $dQ > 0$. The system's Z elements irreversibly convert its mechanical, electrical, or hydraulic energy into heat energy.

To illustrate this irreversible change, consider a perfectly insulated container filled with water. Heat produced within the container cannot leave it. Likewise, heat cannot flow into the container from its surroundings.

A paddle wheel within the container stirs the water. The power supplied by the paddle wheel is the product of its speed and torque. Some of this torque is caused by the resistance of the water to the motion of the paddles; that is, friction. This frictional torque is the effort needed to move the numerous molecules of water that are in the path of the wheel's paddles.

The molecules acquire kinetic energy as they are pushed by the paddles. However, the motion of each water molecule is unpredictable. The molecules move essentially at random as they collide with each other, the walls of the container, and the paddle wheel. We perceive these increasingly energetic collisions as a rise in temperature.

If we do not supply a driving torque, then the resistance of the water causes the paddle wheel to stop turning. All the kinetic energy of the wheel has been used to increase the temperature of the water. The kinetic energy of the wheel is not lost but transformed into the kinetic energy of the water molecules.

The water molecules never (or, more precisely, with vanishingly small probability) collide with a paddle in such a way as to produce a net torque. Instead, they collide with the wheel's paddles in all places, at all times, and from all directions such that the total effect is zero. The kinetic energy of the water molecules cannot do useful mechanical work.

Entropy is a measure of the disordered kinetic state that we have discussed in this example. In a closed system, this disorder cannot be undone and increases with time. [10]

---

[10] Leading some early interlocutors in the field of thermodynamics to speculate about the heat death of the universe. In this future state, the universe (presumed by the speculator to be a closed system) reaches maximum entropy and no useful work can be done. All becomes chaos and disorder.

However, if our system (or model of it) is not closed then disorder can appear to be abated.

Return to our two volumes, one warmer and the other cooler, that exchange heat. If the totality of our observations are made within the warmer volume, then it appears to us that the total entropy decreases. However, our system is not closed. Like Z elements in our mechanical or electrical models, we have left a gap through which energy can escape.

0 and 1 junctions There is a zero junction wherever there is an object with a temperature. Because there is a single temperature and zero junctions do not store energy, it is necessary that the entropy flowing into the junction equal the entropy flowing out of the junction.

A one junction appears where distinct thermal elements exchange energy directly, typically without an intervening temperature or impedance to the flow of heat. One junctions play a less substantial role in thermal systems than they do in other domains. As we shall see, this is a consequence of the first and second laws of thermodynamics.

Basic elements The second law of thermodynamics states that in a closed system the net entropy increases or remains constant. This places rather severe constraints on the elements that can appear in our models. It also produces model elements with forms unlike what we have seen previously. We will assume throughout that pressure and volume remain essentially constant. This assumption is essentially true for solids and liquids in familiar terrestrial circumstances.

The heat capacity C of a substance quantifies the change of temperature that results from the addition or removal of heat energy. It has units of J/K. As heat flows into the substance, the temperate changes at a rate

$$C\dot{T} = P$$

Using the effort e in place of T and the flow f in place of $\dot{S}$, we obtain a nonlinear expression for a thermal C element

$$C\dot{e} = fe$$

Thermal resistance takes remarkable forms because of the first law of thermodynamics. This law states that energy can change form, but it cannot be created or destroyed. In our other engineering domains, lost energy was implicitly converted into heat. Now, that avenue of escape is closed to us.

We will consider two forms of the thermal resistance: one modeling heat transfer by conduction and the other modeling heat transfer by radiation. In each case, the thermal resistor permits heat to move from one body to another. As in the other domains we have studied, the thermal resistor does not store energy, but it does not discard energy. The resistor is merely a conduit through which power flows.

We examine thermal conduction first. It resembles the gyrator and transformer in having two edges:

$$\xrightarrow{1} \mathrm{R} \xrightarrow{2}$$

Conservation of power requires that the heat flowing in equals the heat flowing out, which imposes the constraint

$$e_1 f_1 = e_2 f_2$$

The flow of heat through the element is from the hotter to the colder edge and it is in proportion to the difference of temperature, therefore

$$e_1 - e_2 = Re_1 f_1$$

The thermal resistance *R* to heat transfer by conduction has units of K/W.

The element R has a total entropy that does not decrease, as required by the second law of thermodynamics. The lower limit of temperature permitted by physics as we understand it is zero Kelvin. All temperatures are strictly above this limit. Therefore, $e_1 > 0$ and $e_2 > 0$.

The rate of change of entropy in the resistor is

$$\dot{S}_1 + \dot{S}_2 = f_1 + f_2$$

We want to show that

$$f_1 + f_2 \geq 0$$

Suppose $e_1 \geq e_2$. It follows that $e_1/e_2 \geq 1$ and $f_1 \geq 0$. Conservation of power requires

$$f_2 = \frac{e_1}{e_2} f_1$$

and so

$$f_1 + f_2 = \left(1 + \frac{e_1}{e_2}\right) f_1 \geq 0$$

Thermal convection is the movement of heat energy by the movement of the mass that stores it. A thorough treatment of the subject when fluid and gas flows are involved is much more complicated than can be accommodated in a method for system modeling. However, we can treat the case in which heat is transferred to a body by a fluid or gas flowing over, through, or around that body.

The resistance R in this case is determined by the heat transfer coefficient h of the fluid and the surface area A exposed to the flow. The units of h are $\mathrm{W/(m^2 \cdot K)}$. The thermal resistance to heat transfer by convection is

$$R = \frac{1}{hA}$$

The governing equations are those of the R element.

A variation of the thermal resistor permits us to connect Z elements in other domains to the thermal domain. This ZS element models the transformation of power in the electrical, mechanical, and hydraulic domains into heat. The ZS element has two edges:

$$\xrightarrow{1} \mathrm{ZS} \xrightarrow{2}$$

We take $f_2 = \dot{S}$ and $e_2 = T$. The variables $f_1 = f$, and $e_1 = e$ may come from the mechanical, electrical, or hydraulic domains. The governing equations are

$$\begin{aligned} \dot{S}T &= ef \\ Zf &= e \end{aligned}$$

The thermal F element does not exist. If such a thing were to exist, then it would allow us to store and discharge entropy flow. Indeed, suppose we found this element and that we placed it into the thermal analog of the tank circuit from the chapter on electrical systems. The total entropy in this closed system would rise and fall as the F element absorbs and discharges entropy. We would have found a counterexample to physical laws of thermodynamics.

Radiation Thermal radiation occurs when a heated object emits or absorbs electromagnetic energy. For example, when a hot iron glows, it is radiating heat in the form of visible light. When you stand in the sunshine, you are warmed by electromagnetic energy radiated by the Sun.

The ability of a body to emit and absorb radiated heat is a function of its material composition, surface area, and wavelength of the electromagnetic emission. To model this, we can define a resistance like element with parameter

$$\mathcal{R} = \sigma A \epsilon(\lambda)$$

The Stefan-Boltzmann constant σ is 5.670373×10^{-8} W/(m^2 ·K^4), A is the surface area of the body exposed to radiation, and $\epsilon(\lambda)$ is the emissivity as a function of the wavelength λ. Emissivity is a dimensionless quantity between 0 and 1 and the shape of this function depends on the composition of the radiating body.

With some work, we can see that the units of $\mathcal{R}$ are W/K^4. The Stefan-Boltzmann equation relates temperature and power by

$$P = \mathcal{R}T^4$$

Because $P = \dot{S}T$ we can divide both sides by T to obtain the relation

$$\dot{S} = \mathcal{R}T^3.$$

The thermal radiator has two edges.

$$\xrightarrow{1} \mathcal{R} \xrightarrow{2}$$

The constitutive relations for this element are

$$
\begin{aligned}
e_1 f_1 &= e_2 f_2 \\
f_1 &= \mathcal{R} e_1^3
\end{aligned}
$$

These relations require e_1 to be attached to the zero junction that models the temperature of the radiating body. Temperature is always positive and so this thermal resistor causes entropy to increase.

Heat engines A heat engine can be modeled with an element that has three edges. Although this element does not model the internal operation of the engine, it can describe the engine as a motive force in a larger system. The W element[11] appears in our models as

$$
\begin{array}{c}
\xrightarrow{1} \mathrm{W} \xrightarrow{2} \\
\big\downarrow 3
\end{array}
$$

The heat engine is characterized by an efficiency parameter η. Edges 1 and 2 have thermal variables. Edge 3 connects to the mechanical, electrical, or hydraulic system that the engine supplies with power. The governing equations of the W element are

$$
\begin{aligned}
\dot{S}_1 &= \dot{S}_2 \\
\dot{S}_1 T_1 &= \dot{S}_2 T_2 + e_3 f_3 \\
e_3 f_3 &= \eta \dot{S}_1 T_1
\end{aligned}
$$

The first equation models an engine without internal losses so that the entropy flow entering the engine equals the

[11] W for Work. C for Carnot would have been appropriate, but we have already used C in several contexts! Jean Thoma uses Engine for this element in the monograph *Energy, Entropy, and Information*, IIASA Research Memorandum, June 1977.

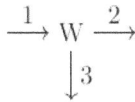

entropy flow leaving the engine. The second equation is conservation of power.

From the second and third equations we may obtain Carnot's formula for the efficiency of a heat engine. Substituting $e_3 f_3$ from the third equation into the second equation we have

$$\dot{S}_1 T_1 = \dot{S}_2 T_2 + \eta \dot{S}_1 T_1$$

Dividing through by $\dot{S}_1$ and recalling that $\dot{S}_1 = \dot{S}_2$ we find

$$T_1 = T_2 + \eta T_1$$

Finally, solving for η we arrive at the familiar form of Carnot's efficiency of a heat engine

$$\eta = 1 - \frac{T_2}{T_1}$$

Drawings to graphs Each temperature appears as a zero junction. If this is an object that can change temperature, then a C element is attached to the zero junction. A source of effort is attached if the temperature is imposed on the system. Two zero junctions that exchange heat by conduction or convection are connected through an R element.

If heat flows by radiation, then the zero junctions are connected by an $\mathcal{R}$ element. If both junctions can emit thermal radiation, then the $\mathcal{R}$ elements appear in pairs. There is one element for heat radiating from body A to body B and the other for heat radiating from body B to body A.

A one junction is used to connect elements from other domains through a ZS element or W element. A one junction will also be used when, for instance, two resistive elements are connected by an ideal thermal conductor.

As always, consistency in your edge directions simplifies the system of equations. Entropy flows from high temperature to low temperature. Therefore, you will want to orient your entropy flows away from hotter elements and toward cooler elements.

Electric heater An electric heating element is a device that uses electrical resistance to produce heat. The mass of the heating element allows it to retain some of its heat even after the electrical power is turned off. These heating elements are common in electric stoves, toaster ovens, and other appliances.

The electrical aspect of this device consists of a voltage source v in series with an electrical resistor. The graph of this electrical system appears below.

$$\begin{array}{ccccc} & & \mathrm{Z} & & \\ & & \uparrow & & \\ \mathrm{SE}: v \to 0 & \longrightarrow & 1 & \longrightarrow & 0 \to \mathrm{SE}: 0 \end{array}$$

To model the thermal aspect of the heater, we replace the Z element with a ZS element. The thermal mass of the resistor is modeled with a C element. The graph appears as such:

$$\begin{array}{ccccc} & & \mathrm{C} & & \\ & & \uparrow & & \\ & & 0 & & \\ & & \uparrow & & \\ & & \mathrm{ZS} & & \\ & & \uparrow & & \\ \mathrm{SE}: v \to 0 & \longrightarrow & 1 & \longrightarrow & 0 \to \mathrm{SE}: 0 \end{array}$$

This model has a single electrical flow (current) i and electrical effort (voltage) v. There is a single temperature T

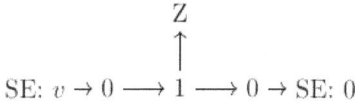

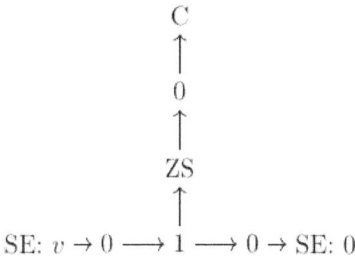

and entropy flow $\dot{S}$. The system of equations for this graph is

$$v = Zi$$
$$iv = T\dot{S}$$
$$\dot{T}C = T\dot{S}$$

This system of equations can be reduced to the single equation

$$\dot{T}C = v^2/Z$$

This model has the curious property that the electrical element never cools! The term v^2 is nonnegative. Therefore, the element heats up when the power is turned on and remains at its temperature when the power is turned off.

Suppose that the electrical element is cooled primarily by blowing cold air over its surface or by conduction into some other material. In both cases, we assume that the heat flows into an environment at an essentially constant temperature T_a. A thermal resistance R models the flow of heat from the resistor into the environment. Augmenting the graph with the source of temperature and thermal resistance gives us the graph

$$
\begin{array}{c}
\text{SE: } T_a \\
\downarrow \\
0 \\
4 \uparrow \\
\text{R} \\
3 \uparrow \quad 2 \\
0 \xrightarrow{\ 2\ } \text{C} \\
1 \uparrow \\
\text{ZS} \\
\uparrow \\
\text{SE: } v \to 0 \longrightarrow 1 \longrightarrow 0 \to \text{SE: } 0
\end{array}
$$

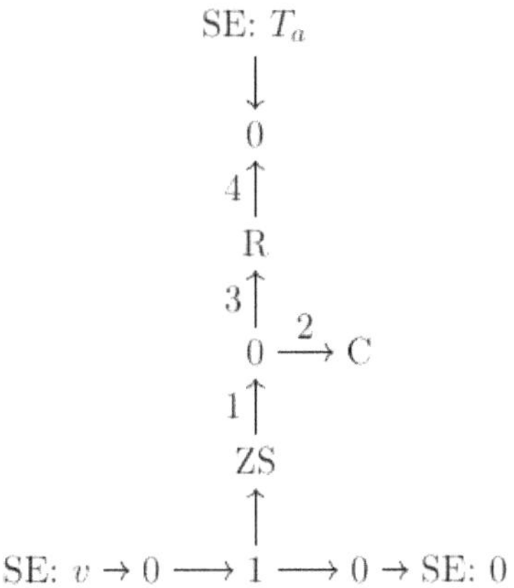

The thermal edges are labeled to distinguish the entropy flows. We keep T as the temperature of the heating element. Notice that $T_4 = T_a$ and

$$T = T_1 = T_2 = T_3$$

The system of equations from this graph is

$$
\begin{aligned}
v &= Zi \\
iv &= T\dot{S}_1 \\
\dot{T}C &= T\dot{S}_2 \\
\dot{S}_1 &= \dot{S}_2 + \dot{S}_3 \\
T - T_a &= R\dot{S}_3 T \\
T\dot{S}_3 &= T_a\dot{S}_4
\end{aligned}
$$

Climate change The Sun radiates heat in the form of visible light. The Earth's atmosphere is very nearly transparent to visible light. Consequently, the surface of the Earth that is exposed to sunlight absorbs much of this radiated heat.

On the Earth's night side, heat that was acquired on the day side radiates back into space as infrared light. Carbon dioxide in the atmosphere is opaque to infrared light, thereby trapping some of this radiated heat. Indeed, this trapping of heat energy by carbon dioxide is necessary to maintain the pleasant temperatures upon which life depends.

However, the atmosphere becomes increasingly opaque to infrared light as the amount of carbon dioxide increases. An increasingly opaque atmosphere causes more of the heat absorbed on the day side of the planet to be retained on the night side. The consequence is an increasing temperature; that is, global warming.

A very simple model of this dynamic can be constructed with four components. The Sun is modeled by a source of power. The Sun delivers its heat power to Earth, which we model as a thermal C element. On the night side of the planet, carbon dioxide in the atmosphere offers resistance $\mathcal{R}$ to heat radiating away from the surface. The emissivity of the Earth through the atmosphere at infrared frequencies is a function of its carbon dioxide content c.

The heat is radiated into space, which has an approximately constant temperature. Space is modeled by a source of temperature that absorbs entropy from the C element by way of the $\mathcal{R}$ element. The graph of the model appears as such:

$$\begin{array}{ccccccccc} & & \mathrm{C} & & & & & & \\ & & 2\uparrow & & & & & & \\ \mathrm{SP} & \xrightarrow[1]{} & 0 & \xrightarrow{3} & \mathcal{R} & \xrightarrow{4} & 0 & \xleftarrow[5]{} & \mathrm{SE} \end{array}$$

Let $T = e_1 = e_2 = e_3$ be the temperature of the Earth and $e_5 = e_4 = T_s$ be the temperature of space. The system of equations is

$$\dot{T}C = T\dot{S}_2$$
$$\dot{S}_1 = \dot{S}_2 + \dot{S}_3$$
$$\dot{S}_3 = \mathcal{R}(c)T^3$$
$$\dot{S}_3 T = \dot{S}_4 T_s$$

To simulate this model, we change units so that T is the fractional change in temperature relative to some pre-industrial reference temperature. This allows us to use $T = 1$ for our initial temperature, and, for example, $T = 1.1$ is a 10% increase in temperature above the reference point.

Our unit of power is the total power delivered by the Sun to the Earth. Therefore, the input power $T\dot{S}_1 = 1$. The temperature of space is 2.7 K. We use 288 K for our pre-industrial reference temperature, making $T_s = 2.7/288 \approx 0.0094$.

In a similar fashion, $c = 1$ is the carbon dioxide content of the pre-industrial atmosphere. We model the emissivity of the Earth as a function of c such that

$$\mathcal{R}(c) = \mathcal{R}_0/c$$

The constant $\mathcal{R}_0$ is to be determined. If we assume the preindustrial temperature of the Earth was approximately stable, then

$$\dot{T} = 0 \text{ therefore}$$
$$\dot{S}_2 = 0\,,\ \dot{S}_1 = \dot{S}_3\,, \text{ and}$$
$$\mathcal{R}_0 T^4 = \dot{S}_1 T$$

Because $T = 1$ and $\dot{S}_1 T = 1$, we find $\mathcal{R}_0 = 1$.

Table 1. Carbon dioxide relative to 1850 (from gml.noaa.gov)

years since 1850	c
0	1
133	1.214
165	1.429
175	1.5

The carbon dioxide concentration in the atmosphere has increased by approximately 50% over pre-industrial levels. The rate of increase has grown with the accelerating rate of industrial development. Industrial growth has followed something like a Weibull function during that period. By fitting the handful of data points in Table 1 to the Weibull like function

$$c(t) = \exp((at)^b)$$

we obtain $a = 0.00408$ and $b = 2.659$.

The average global temperature has risen by approximately 1.1 K since pre-industrial times. For our model to match this increase in temperature, we expect $T(175) = 289.9/288 \approx 1.0038$. By setting $C = 4.5 \times 10^3$ our model reaches this temperature, this choice for C being made by trial and error with a simulation.

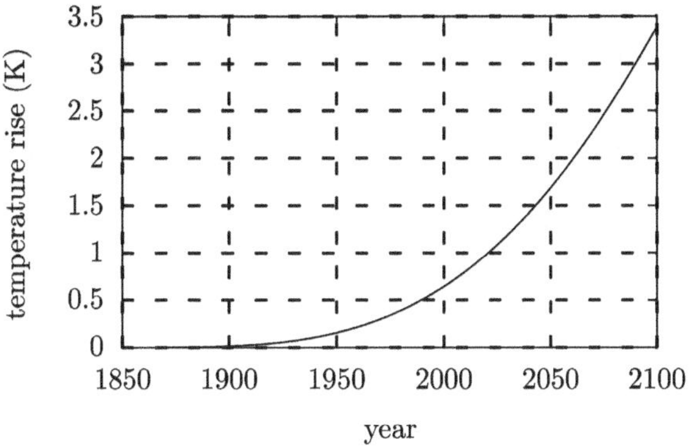

Figure 15. Temperature rise above the pre-industrial average, calculated as (1-T)288 K for each year since 1850.

The temperatures produced by this model are shown in Figure 15. Just as the IPCC does, we have simulated our model forward in time to the year 2100. The temperature reaches $T \approx 1.012$, which is about 3.5 K above the pre-industrial reference.

Our model is crude. Nevertheless, our result corresponds roughly to IPCC Scenario 4: the 'business as usual' emissions scenario. This is not too surprising. We have merely projected forward using historical data and some understanding of thermodynamics to reach a conclusion that has been understood since the 1950s.

Steam locomotive Recalling our model of a train (see the chapter on transformers and gyrators), let us exchange the electric motor for an (idealized) steam engine. The train engineer provides heat to the engine with a coal or wood burning fire. The heat of the fire creates steam from the water in a boiler. We model the heated water and steam as a source of temperature T_f.

The pressure of the steam drives the pistons connected to the wheels. We model this aspect of the engine with a W element. Some of the steam provides mechanical power; this power flows through edge 5 of the W element in our graph. Edge 8 of the W element is where the engine expels the unused steam (that is, waste heat) into the atmosphere at temperature T_a.

A mechanism for controlling the propulsive force is to restrict the flow of steam from the boiler to the pistons. We model the restriction of heat flow with an R element. If this element is used to control the speed s then its parameter is a function $R(s)$. Of course, we hope there is also a mechanical brake, but we do not include it in our model.

The simplified graph of the steam locomotive is shown below.

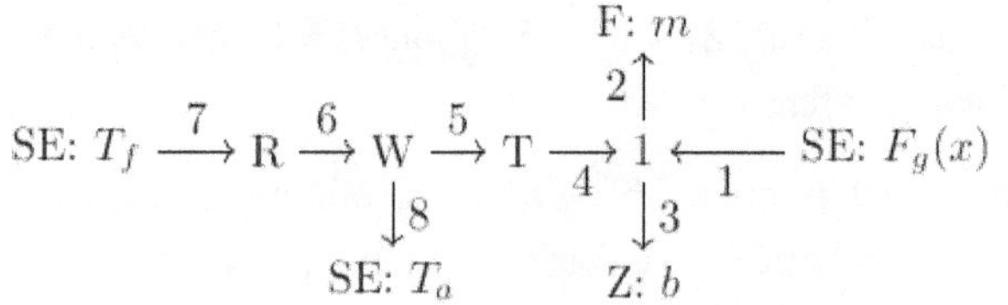

The system of equations for this graph is

$$T_f - T_6 = R(s)T_6\dot{S}_6$$
$$T_f\dot{S}_7 = T_6\dot{S}_6$$
$$\dot{S}_6 = \dot{S}_8$$
$$\dot{S}_6 T_6 = e_5 f_5 + \dot{S}_8 T_a$$
$$e_5 f_5 = \eta \dot{S}_6 T_6$$
$$e_4 T = e_5$$
$$f_5 T = f_4$$
$$\dot{f}_2 = (e_4 - bf_2 + F_g(x))/m$$
$$s = \dot{f}_2$$
$$\dot{x} = f_2$$

Exercises

Ex. 8.1 Two thermal insulators are sandwiched between a hot and a cold source of temperature, as shown below. Draw the graph for this system. Show that the sandwiched insulators R_1 and R_2 are equivalent to a single insulator R with parameter $R = R_1 + R_2$.

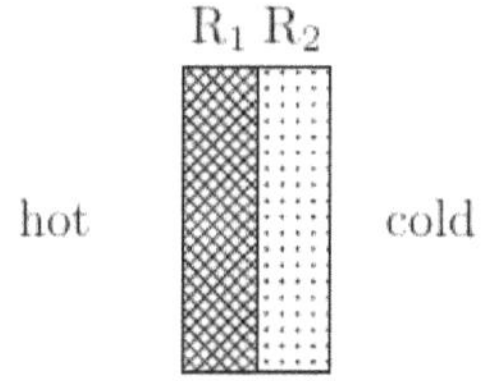

Ex. 8.2 A baseboard and drill is used to start fires by friction. The fire maker applies pressure to the drill as it turns against the baseboard. This creates fine particles that ignite to create a burning ember. Build a model of this system. You may assume that the rotational friction between the drill and the baseboard is a function KF where K is a constant and $F > 0$ is the force indicated in the diagram. Include the angular speed of the drill, the thermal capacitance of the board, and the creation of heat by friction. Heat is lost by conduction

through the baseboard away from the drill hole. You may assume that the part of the baseboard away from the drill hole is at ambient temperature.

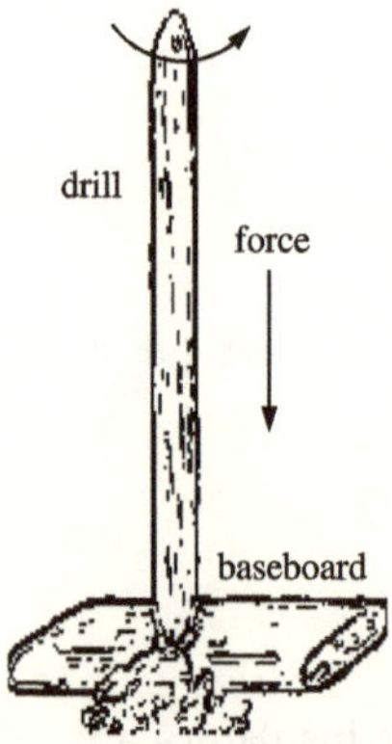

Ex. 8.3 Suppose that the ember ignites when the temperature at the drill hole reaches a temperature T. How much energy has been expended by the fire maker when ignition occurs? What can you conclude about the types of wood that are preferred (that is, what are our preferences for R and C)?

Ex. 8.4 Consider a long, thermally conductive rod. Divide the rod into segments of length h. Each segment has a thermal capacitance Ch. The conduction of heat between segments encounters a thermal resistance Rh. The left end of the rod is in a hot bath (source of effort) and the right end is in a cold bath. The rod is perfectly insulated everywhere else. Draw a graph for the rod divided into two segments. What partial differential equation does this converge to as the number of segments goes to infinity (that is, at h goes to zero)? Hint: Write the equation for $\dot{T}$ at one of the elements as a function of T at the element and its neighboring elements.

Ex. 8.5 Create the graph of the system shown below. The two identical thermal masses (drawn as circles) are separated by an insulator (the hatched box) from each other and the surrounding, ambient temperate. Heat transfer is by conduction.

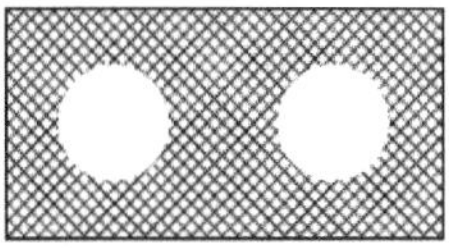

Ex. 8.6 A pseudo-graph for thermal systems can be constructed by using power for the flow variable. This is a psuedo-graph in the sense that the product of effort T and power P is not power. Derive the constitutive relations of the one and zero junctions and the C, R, and W elements for the psuedo-graph.

Ex. 8.7 A small building has two rooms that may be modeled by C elements. Heat transfer between the rooms can be modeled as an R element. The walls of the building insulate it from the outside air temperature. A heat pump, which may be modeled as a W element, supplies air conditioning to one of the rooms. Build a graph for this system using the psuedo-thermal elements created in the previous exercise.

9. The phasor

The phasor is a tool for understanding how a system responds to a periodic stimulus. This problem appears frequently in engineering. Signal filters, such as those used in audio systems and radars, are designed to dampen noise at particular frequencies. Electric circuits often carry periodic power signals. For example, the ubiquitous power outlet supplies a 50 Hz or 60 Hz periodic signal, depending on the power standards where you live. Vibrations, such as those caused by seismic tremors, are periodic forces that must be considered when designing structures.

The frequency response is usually examined for linear systems, and we will focus on that case here. It can be extended, with some effort, to nonlinear systems, but this is uncommon. Numerical simulations are more often used to analyze nonlinear systems. However, there are important exceptions, particularly when looking at electrical power systems, where a nonlinear problem describes how power flows through the electrical network.

Phasors A phasor is a complex function of time

$$u(t) = r\cos(2\pi\nu t + \phi) + ir\sin(2\pi\nu t + \phi)$$

The signal $u(t)$ has a period $1/\nu$, phase angle ϕ, and amplitude r. The real part of $u(t)$ is just that, real. We can measure it with an instrument. The imaginary part is a clever addition by Steinmetz to aid analysis.[12]

The essential idea is to interpret $u(t)$ as a real signal that exists on a circle. The radius r of the circle is the amplitude of the signal. If we take a measurement at some time, then we will obtain a real component x (on the x-axis of our circle) and an imaginary component y (on the y-axis of our circle). Clearly,

$$x^2 + y^2 = r^2$$

and the two components form an angle

$$\tan^{-1}(y/x) = \phi$$

The vector (x, y) rotates with a frequency ν passing through 2π radians every $1/\nu$ units of time. Suppose that we can only see x, the real part of our complex signal. Then we perceive a signal

$$x(t) = r\cos(2\pi\nu t + \phi)$$

---

[12] The idea of a phasor was presented over a century ago in the context of electrical engineering; see Steinmetz's paper *Complex Quantities and Their Use in Electrical Engineering*, in Proceedings of the International Electrical Congress Held in the City of Chicago, American Institute of Electrical Engineers, pp. 33–74, August 21st to 25th, 1893. The Fourier series was known in the 1800s, but separately and without obvious application to the problems faced by engineers. A history of the phasor is presented by Antônio Araújo and Danny Tonidandel in *Steinmetz and the Concept of Phasor: A Forgotten Story*, Journal of Control, Automation and Electrical Systems, #24, published 2013.

Nonetheless, the invisible (to our measurement device) $y-$ axis allows our calculations to involve simple algebra with complex numbers instead of solutions to differential equations. This is the value of the phasor.

Our signal is periodic with a fixed frequency v. Therefore, it is completely characterized by r and ϕ. Choose a time t_0 such that we take measurements precisely at moments $t_0 + n/v$ where n is an integer. These measurements have the form

$$u_n = r\cos(2\pi v(t_0 + n/v) + \phi) + ir\sin(2\pi v(t_0 + n/v) + \phi)$$

Recall that

$$\cos(2\pi n + \theta) = \cos(\theta)$$
$$\sin(2\pi n + \theta) = \sin(\theta)$$

Consequently,

$$u_n = r\cos(2\pi v t_0 + \phi) + ir\sin(2\pi v t_0 + \phi)$$

Our choice of t_0 produces a change in the perceived phase angle ϕ. However, t_0 is the reference point for every measurement in our system, and so this phase shift occurs in all of our signals. A convenient choice is $t_0 = 0$. With this choice, u_n is the complex number

$$u = r\cos(\phi) + ir\sin(\phi)$$

The set of phasors with frequency v is closed under addition. To see this, we use Euler's formulae to write the phasor $u(t)$ as the complex exponential

$$u(t) = re^{i2\pi vt}e^{i\phi}$$

Consider two phasors, one with amplitude r_1 and angle ϕ_1 and the other with amplitude r_2 and angle ϕ_2. Their sum

$$r_1 e^{i2\pi\nu t} e^{i\phi_1} + r_2 e^{i2\pi\nu t} e^{i\phi_2} = e^{i2\pi\nu t}\left(r_1 e^{i\phi_1} + r_2 e^{i\phi_2}\right)$$

is a phasor with frequency ν. Setting $t = t_0 = 0$, we find that adding phasors is indistinguishable from adding their expressions as complex numbers.

The product of phasor $u(t)$ and complex number $a + ib$ is

$$(a + ib) r e^{i2\pi\nu t} e^{i\phi} = a r e^{i2\pi\nu t} e^{i\phi} + i b r e^{i2\pi\nu t} e^{i\phi}$$

The first term in this expression is a phasor with frequency ν. The second term is

$$\begin{aligned} i b r e^{i2\pi\nu t} e^{i\phi} &= i b r \cos(2\pi\nu t + \phi) - b r \sin(2\pi\nu t + \phi) \\ &= b r \cos\left(2\pi\nu + \phi + \frac{\pi}{2}\right) + \\ &\qquad i b r \sin(2\pi\nu t + \phi + \pi/2) \\ &= b r e^{i2\pi\nu t} e^{i(\phi + \pi/2)} \end{aligned}$$

This term is also a phasor with frequency ν, amplitude br, and phase $\phi + \pi/2$. Therefore, the product of a phasor and a complex number is also a phasor.

The time derivative of a phasor is

$$\begin{aligned} \dot{u}(t) &= (i2\pi\nu) r e^{i\phi} e^{i2\pi\nu t} \\ &= i2\pi\nu u(t) \end{aligned}$$

and we see that the time derivative is also a phasor with frequency ν.

To summarize these results, let $u(t)$ be the time domain signal with frequency ν and phase ϕ. Let $\hat{u}(\nu)$ be its corresponding phasor. The phasor transform $u(t) \leftrightarrow \hat{u}(\nu)$ has the following properties:

$$\hat{u}(\nu) = re^{i\phi} = r(\cos(\phi) + i\sin(\phi))$$
$$u(t) + v(t) \leftrightarrow \hat{u}(\nu) + \hat{v}(\nu)$$
$$cu(t) \leftrightarrow c\hat{u}(\nu)\text{ , }c\text{ a complex constant}$$
$$\dot{u}(t) \leftrightarrow i2\pi\nu\hat{u}(\nu)$$

Model elements Closure under addition leaves zero and one junctions unchanged in the phasor domain. The Z, E, and F elements in the phasor domain have the form

$$Z\hat{f} = \hat{e}\text{, for the Z element}$$
$$i2\pi\nu E\hat{e} = \hat{f}\text{, for the E element}$$
$$i2\pi\nu F\hat{f} = \hat{e}\text{, for the F element}$$

These relations reveal the utility of the phasor transform. Any analysis will involve only Z-like elements, with the E and F elements having imaginary Z-like parameters.

Rules for series and parallel composition of elements follow immediately. For example, an E element, F element, and Z element in series can be replaced by a single Z-like element. The parameter Z_{eq} of this single element is

$$Z_{\mathrm{eq}} = Z + i2\pi\nu F + \frac{1}{i2\pi\nu E}$$

If the E, F, and Z element are in parallel, then the equivalent, Z-like element has the parameter

$$\frac{1}{Z_{\mathrm{eq}}} = \frac{1}{Z} + \frac{1}{i2\pi\nu F} + i2\pi\nu E$$

Power Power does not fit neatly into the scheme of phasors. Multiplying an effort with angle ϕ_1 and flow with angle ϕ_2 we find

$$e^{i2\pi\nu t}e^{i\phi_1}e^{i2\pi\nu t}e^{i\phi_2} = e^{i(4\pi\nu t+\phi_1+\phi_2)}$$

and the resulting power has a frequency 2ν rather than ν.

Instead, let us return to the real parts of the effort and flow so that

$$e(t) = A\cos(2\pi\nu t + \phi_1) \text{ and}$$
$$f(t) = B\cos(2\pi\nu t + \phi_2)$$

Taking their product and using the heterodyne identity from trigonometry we find

$$P(t) = e(t)f(t) = \frac{1}{2}AB(\cos(4\pi\nu t + \phi_1 + \phi_2) + \cos(\phi_1 - \phi_2))$$

The average power over a period $1/\nu$ is

$$\nu\int_0^{1/\nu} P(t)\,dt = \frac{1}{2}AB\cos(\phi_1 - \phi_2)$$

Once again, let us interpret this as a measurement of a function on a circle of radius $AB/2$ forming an angle $\phi_1 - \phi_2$ relative to the real (or x −) axis. This interpretation gives average power the form

$$\hat{P} = \frac{1}{2}AB\cos(\phi_1 - \phi_2) + i\frac{1}{2}AB\sin(\phi_1 - \phi_2)$$

The average power phasor $\hat{P}$ is the product of the effort phasor $\hat{e}$ and the complex conjugate of the flow phasor $\hat{f}^*$

$$\hat{P} = \frac{1}{2}\hat{e}\hat{f}^* = \frac{1}{2}ABe^{i2\pi\nu t}e^{i\phi_1}e^{-i2\pi\nu t}e^{-i\phi_2}$$
$$= \frac{1}{2}ABe^{i(\phi_1 - \phi_2)}$$

The real part of the complex power is discernible by the heat, motion, or other work that it does. For example, the Z element dissipates power as heat. Its average power

$$\frac{1}{2}Z\hat{f}\hat{f}^* = \frac{1}{2}ZB^2$$

is real and B is the amplitude of the flow phasor.

The imaginary part of the complex power is contained in the system's energy storing elements. For example, the E element stores and discharges power; it does not dissipate power. The purely imaginary power at the E element is

$$\begin{aligned}\frac{1}{2}(i2\pi\nu E\hat{e})^*\hat{e} &= A^2\pi\nu E e^{i\phi_1}(ie^{i\phi_1})^* \\ &= A^2\pi\nu E e^{i\phi_1}e^{-i(\phi_1+\pi/2)} \\ &= A^2\pi\nu E e^{-i\pi/2} \\ &= -iA^2\pi\nu E\end{aligned}$$

where A and ϕ_1 are the amplitude and phase of the effort phasor. The imaginary part of power is called reactive power by electrical engineers.

Root mean square phasors We have been discussing phasors with amplitudes equal to the maximum amplitude of the related sinusoid. Often in practice, the root mean square amplitude of the sinusoid is used instead. The root mean square amplitude of the effort is

$$e_{\mathrm{rms}} = \sqrt{\nu \int_0^{1/\nu} A^2 \cos^2(2\pi\nu t + \phi_1)\, dt}$$

$$= A \sqrt{\nu \int_0^{1/\nu} 1 + \cos(4\pi\nu t + 2\phi_1)\, dt}$$

$$= \frac{A}{\sqrt{2}}$$

and likewise for the flow. Consequently,

$$\hat{e}_{\mathrm{rms}} = \frac{1}{\sqrt{2}}\hat{e} \text{ and } \hat{f}_{\mathrm{rms}} = \frac{1}{\sqrt{2}}\hat{f}$$

and power is

$$\hat{P} = \hat{e}_{\mathrm{rms}}\hat{f}^*_{\mathrm{rms}} = \frac{1}{2}\hat{e}\hat{f}^*$$

Filters A filter is a device that accepts a signal that is the sum of several frequencies, removes the frequencies that are considered noise or are otherwise uninteresting, and passes through the frequencies of interest. The phasor is a tool for studying the behavior of systems that act as filters.

A filter is classified by its action on the signal passing through it. A low pass filter permits frequencies below some threshold and attenuates frequencies above it. A high pass filter has the opposite effect. A band pass filter permits frequencies within a range of frequences and attenuates frequencies outside that range. A notch filter attenuates frequencies within a range and permits frequencies outside that range.

Consider the graph fragment

$$\begin{array}{ccccc} & & \mathrm{F} & & \\ & & 2\big\uparrow & & \\ \mathrm{SE} & \xrightarrow[1]{} & 1 & \xrightarrow[3]{} & \mathrm{Z} \end{array}$$

The source of effort e_1 is the input to this filter. The output is the effort e_2 across the F element. If the input is a phasor, then the efforts and the single flow are related by

$$\hat{e}_1 = \hat{e}_2 + \hat{e}_3$$
$$i2\pi\nu F\hat{f} = \hat{e}_2$$
$$Z\hat{f} = \hat{e}_3$$

Rearranging the equations we find that the input to output relationship of this filter is

$$\hat{e}_1 = \left(1 + \frac{Z}{i2\pi\nu F}\right)\hat{e}_2$$

The gain G of the filter is the ratio of the amplitudes of the output and input. After some simplification and rearranging, we find that

$$G = \left|\frac{\hat{e}_2}{\hat{e}_1}\right| = \frac{2\pi\nu F/Z}{\sqrt{4\pi^2\nu^2F^2/Z^2 + 1}}$$

The ratio F/Z is a design parameter for this filter. By choosing it, we can determine which frequencies are allowed to pass and which are absorbed.

The frequency response of this system is shown in Figure 16 for several choices of F/Z. Its role as a high pass filter is clear. Low frequency signals are severely attenuated while high frequency signals are allowed to pass. The choice of F/Z decides where the attenuation region begins.

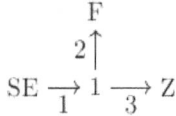

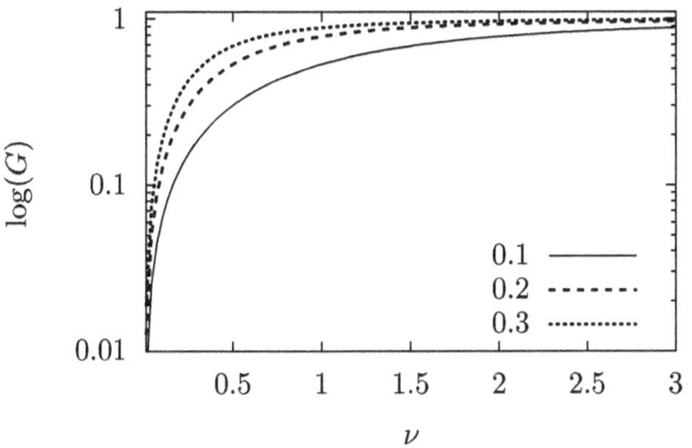

Figure 16. Filter gain as a function of frequency v for F/Z equal to 0.1, 0.2, and 0.3.

Power flow The power flow problem appears when we study electrical power networks. The solution to a power flow problem tells us the phases and amplitudes of the efforts and powers when the system is in its nominal operating condition. We will set up an example of the problem, which produces a system of nonlinear equations that must be solved using an appropriate numerical method.

To begin, let us look back to the transmission line model in the chapter on electric circuits. We replace the Z: R_2 element and the SF element with either a source of power or a source of effort. The flow f_0 (current i_0 in the circuit diagram) enters the transmission line and the flow f_2 (current i_4 in the circuit diagram) leaves the line. We use S_1 and S_2 for the undecided source elements. The graph of the line is

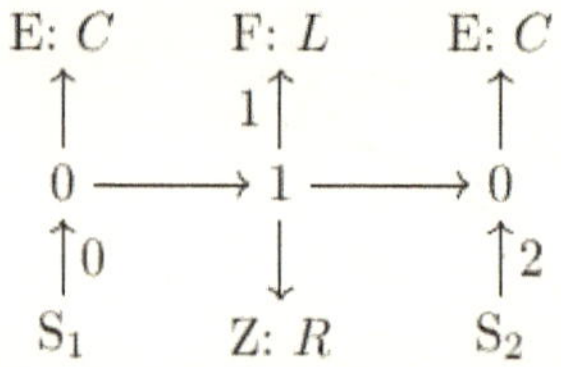

Under normal conditions, an electrical power system carries voltages and currents having a fixed frequency of 60 Hz (in the United States) or 50 Hz (in much of the rest of the world). Therefore, this graph's system of equations when written for analysis at the nominal operating frequency is

$$
\begin{aligned}
i2\pi\nu C\hat{e}_0 &= \hat{f}_0 - \hat{f}_1 \\
i2\pi\nu L\hat{f}_1 &= \hat{e}_0 - \hat{e}_2 - R\hat{f}_1 \\
i2\pi\nu C\hat{e}_2 &= \hat{f}_1 + \hat{f}_2
\end{aligned}
$$

This system of three equations has five variables: two efforts and three flows. One of four possible combinations of sources completes the set of equations.

If we are given two sources of effort, then what must be solved for are the three flows. If we are given two sources of power, then the missing equations become

$$
\begin{aligned}
\hat{p}_0 &= \hat{e}_0\hat{f}_0^* \quad \text{power entering the line on the left} \\
\hat{p}_2 &= \hat{e}_2\hat{f}_2^* \quad \text{power entering the line on the right}
\end{aligned}
$$

A mixture of sources of effort and sources of power are also adequate to create equal numbers of equations and unknown variables.

To define the general power flow problem, suppose that several transmission lines are connected together to form a network. Label the efforts of the zero junctions as e_0, e_1, ..., e_n. Label the flows adjacent to a source (of effort or power) as $f_0, f_1, \ldots, f_n$. Label the flows into the transmission lines according to the zero junctions at each end of the line so that f_{12} is the flow into the one junction that separates efforts e_1 and e_2. Observe that $f_{jk} = -f_{kj}$.

On each line from j to k, the F and Z elements are in series and can be replaced with a single, complex impedance

$$z_{jk} = i2\pi\nu L_{jk} + R_{jk}$$

It will be convenient to define the *admittance* of this line as

$$y_{jk} = 1/z_{jk}$$

Now consider the flow f_k of the source element attached to the kth zero junction. Also attached to this zero junction is a capacitance $i2\pi\nu C_k$. Where several lines have come together, their individual E elements are in parallel and can be replaced with a single E element whose parameter is their sum.

If J is the set of lines incident to the kth zero junction, then the flow f_k out of the source and into the network must satisfy

$$\begin{aligned}\hat{f}_k &= \hat{e}_k i2\pi\nu C_k + \sum_{j\in J} \hat{f}_{kj} \\ &= \hat{e}_k i2\pi\nu C_k + \sum_{j\in J} (\hat{e}_k - \hat{e}_j) y_{jk}\end{aligned}$$

Adding the additional requirements for the power injected at each zero junction gives

$$\hat{p}_k = \hat{e}_k \hat{f}_k^*$$

There are $3n$ variables in these equations: n power variables, n effort variables, and n flow variables. The sources supply us with n of these variables. There are $2n$ equations: a sum of flows at each zero junction and the power injected into each zero junction by the adjacent source. Consequently, there are $2n$ unknowns in $2n$ equations and the system is solvable.[13]

Vibration Consider the automobile body and its suspension described in the chapter on mechanics. We wish to know how intensely the body of the vehicle is shaken as a function of the frequency content produced by a bumpy stretch of road. The vehicle's equations of motion transformed into the phasor domain are

$$\begin{aligned}
&i2\pi\nu\hat{e}_2 = (\hat{v} - \hat{f}_1)/k_2 \\
&i2\pi\nu\hat{e}_3 = \hat{f}_3/k_1 \\
&i2\pi\nu\hat{f}_1 = (\hat{e}_2 - \hat{e}_3 - \hat{f}_3 b)/m_2 \\
&\hat{f}_3 = \hat{f}_1 - \hat{f}_4 \\
&i2\pi\nu\hat{f}_4 = (\hat{e}_3 + \hat{f}_3 b - m_1 g)/m_1
\end{aligned}$$

The periodic motion imposed on the tire by the road is $\hat{v}$. The consequent shaking of the car body is $\hat{f}_4$. The equations of motion can be written in matrix form as

[13] Unless, of course, the system is degenerate. Power flow calculations for very large networks necessarily use estimates of parameter and variable values. If these estimates are far enough from the correct values or produce a nearly degenerate system of equations, then we will encounter severe difficulties in calculating a numerical solution.

$$\begin{bmatrix} i2\pi\nu & 0 & 1/k_2 & 0 & 0 \\ 0 & i2\pi\nu & 0 & -1/k_1 & 0 \\ -1/m_2 & 1/m_2 & i2\pi\nu & b/m_2 & 0 \\ 0 & 0 & -1 & 0 & 1 \\ 0 & -1/m_1 & 0 & -b/m_1 & i2\pi\nu \end{bmatrix} \begin{bmatrix} \hat{e}_2 \\ \hat{e}_3 \\ \hat{f}_1 \\ \hat{f}_3 \\ \hat{f}_4 \end{bmatrix} = \begin{bmatrix} \hat{v}/k_2 \\ 0 \\ 0 \\ 0 \\ -g \end{bmatrix}$$

We can assign numbers to the parameters and plot the frequency response of this system.[14] A small car will have a mass of about $m_1 = 1{,}100$ kg. The stiffness of the rubber tire might be around $1/k_2 = 200{,}000$ N/m with a mass $m_2 = 10$ kg. Reasonable suspension values for a small car are $k_1 = 1/25{,}000$ N/m and $b = 2{,}000$ N·s/m.

Figure 17 plots $|\hat{f}_4|$ as a function of the frequency ν in response to $\hat{v} = 1$ m/s. The maximum response is obtained at a frequency just above 2 Hz. The model is linear and, therefore, the magnitude of the response scales with the magnitude of $\hat{v}$. A 1 m/s bump in the road is unrealistic and the response must be scaled to match the expected conditions.

Suppose that we want to place bumps in a parking lot surface to discourage unsafe speeds, perhaps encouraging an upper limit of 10 miles per hour, which is 2.24 m/s. At that speed, we wish to induce the maximum possible vibration of the vehicle. This can be accomplished by placing the bumps at a specific interval. Distance d, frequency ν, and speed v are related by

[14] These data are from J.A. Calvo, V. Diaz, and J.L. San Roman, *Establishing inspection criteria to verify the dynamic behaviour of the vehicle suspension system by a platform vibrating test bench*, International Journal of Vehicle Design, vol. 38, no. 4, pages 290-306, 22 August 2005.

$$d = v/\nu$$

The appropriate distance between speed bumps is $2.24/2 = 1.12$ m.

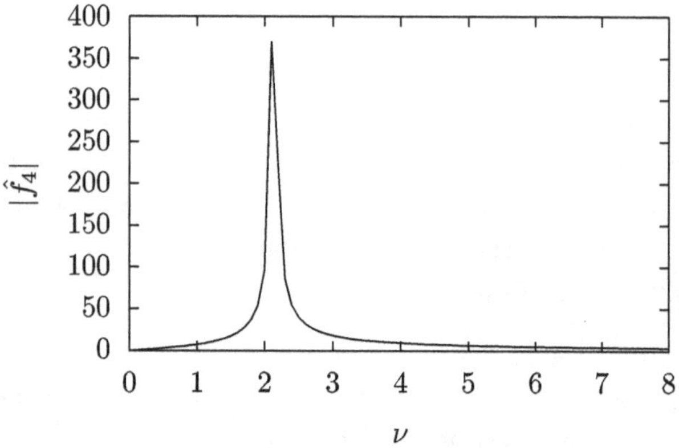

Figure 17. Velocity of the passenger cab, in units of meters per second, in response to a 1 meter per second velocity imposed by the road surface on the tire. The units of ν are Hertz.

Mixing time and phasor domains There are some systems in which the difference in the time scales of dynamic behaviors is so significant that one part can be modeled in the frequency domain and another in the time domain. A notable example is the analysis of generator stability in an electrical power system.

For generator stability studies, the primary concern is the rotational speeds of the generators. These must remain within strict limits. The dynamic behavior of the generators occurs in seconds. The generators' primary role, as seen from the network, is to act as a time varying source of power at a frequency very close to 60 (in the United States) or 50 (in many other parts of the world) Hz.

The electrical system that connects the generators to each other and the electrical customers has dynamic behavior that occurs over milliseconds. Because our interest is chiefly in the generator dynamics and this difference in time scales is so significant, the electrical parts of the system are modeled as though they were affected only by a pure sinusoidal source of power. Any electrical dynamics are assumed to resolve themselves so quickly and to be so insignificant in relation to the generator motions that they can be safely ignored.

We have seen how the electrical network can be modeled in the phasor domain. Let us now consider how a time domain model of the source of power can be constructed. The generators have a nominal speed w_0. The actual speed at any moment is some small deviation ω from the nominal ω_0. The rotating turbine in the generator has angular mass I.

This turbine is driven by a torque τ_1. This torque is produced by steam at high pressure that flows over blades attached to the turbine. The flow of steam is regulated by a controller, in our case a simple proportional-derivative control with parameters $k_1 > 0$ and $k_2 > 0$ that seeks to maintain ω close to zero.

Opposing this supplied torque is another torque τ_2 that originates with the power demanded by the electrical system. The mechanical motion of the generator about its nominal speed ω_0 is described by

$$
\begin{aligned}
\dot{\tau}_1 &= -k_1\omega - k_2\dot{\omega} \\
\dot{\omega} &= (\tau_1 - \tau_2)/I \\
\dot{\theta} &= \omega
\end{aligned}
$$

The angular displacement θ of the turbine is measured relative to the reference angle of the network's phasor model.

It will be significant when we tie the generator model to the network equations.

The generator supplies power to the network with a motor that is driven by the turbine. The motor has three phases a, b, and c with voltages e_a, e_b, e_c and currents f_a, f_b, and f_c. To connect the generator equations with the phasor representation of the electrical network requires $\omega \approx 0$ at all times.

In fact, the generator cannot operate unless this condition is nearly true. Therefore, when considering its connection to the electrical system, we assume $\omega \approx 0$. The power demanded by each phase of the motor is

$$\begin{aligned} p_a &= e_a f_a \\ p_b &= e_b f_b \\ p_c &= e_c f_c \end{aligned}$$

The total power demanded from the generator is

$$\omega_0 \tau_2 = f_a e_a + f_b e_b + f_c e_c$$

The construction of the motor that is driven by the turbine is such that the voltage and current appearing on each phase is a sinusoidal signal in the form

$$\begin{aligned} e_a &= v\cos(\omega_0 t + \theta) \\ e_b &= v\cos(\omega_0 t + \theta + 2\pi/3) \\ e_c &= v\cos(\omega_0 t + \theta - 2\pi/3) \\ f_a &= f\cos(\omega_0 t + \phi) \\ f_b &= f\cos(\omega_0 t + \phi + 2\pi/3) \\ f_c &= f\cos(\omega_0 t + \phi - 2\pi/3) \end{aligned}$$

The angle θ is the angular displacement of the turbine in its dynamic equations. The voltage amplitude v is a constant.

Angle ϕ and flow amplitude f are electrical variables in the network.

In our model, the phases a, b, and c are identical except for their separation by $2\pi/3$ radians. Let us consider just phase a for the moment. With ω near zero, the effort and flow are approximately the phasors

$$\hat{e}_a = ve^{i\theta}e^{i2\pi\omega_0 t}$$
$$\hat{f}_a = fe^{i\phi}e^{i2\pi\omega_0 t}$$

Using this approximation, the generator can be connected to a phasor model of the transmission system. The generator imposes $\hat{e}_a$, $\hat{e}_b$, and $\hat{e}_c$. The resulting $\hat{f}_a$, $\hat{f}_b$, and $\hat{f}_c$ decide the torque τ_2 to which the generator must respond.

To illustrate the resulting dynamics, let us write the full system of equations for a single generator attached to a single, three phase load. Because each phase is identical, we need only consider the single phase a when writing the electrical equations.

The electrical variables for phase a at the generator side of the transmission line are $\hat{f}_0$, $\hat{e}_0$, and $\hat{p}_0$. At the load side of the line these are $\hat{f}_2$, $\hat{e}_2$, and $\hat{p}_2$. The set of electrical equations (copied from our previous example of the transmission line) are

$$i2\pi\nu C\hat{e}_0 = \hat{f}_0 - \hat{f}_1$$
$$i2\pi\nu L\hat{f}_1 = \hat{e}_0 - \hat{e}_2 - R\hat{f}_1$$
$$i2\pi\nu C\hat{e}_2 = \hat{f}_1 + \hat{f}_2$$
$$\hat{p}_0 = \hat{e}_0\hat{f}_0^*$$
$$\hat{p}_2 = \hat{e}_2\hat{f}_2^*$$

The coupling of mechanical and electrical elements is

$$|e_0| = v$$
$$\angle \hat{e}_0 = \theta$$
$$\omega_0 \tau_2 = p_a + p_b + p_c$$

To solve the system of equations, we take the power $\hat{p}_2$ demanded at the load to be constant.

A numerical example will illustrate the behavior of this system. For this example, let us assume a 100 km long transmission line with $C = 0.03\ \mu$F/km, $L = 0.001$ H/km, and $R = 0.06\ \Omega$/km. The angular mass of the turbine is 20,000 kg·m^2 and its nominal speed is $\omega_0 = 120\pi$ rad/s (60 Hz). The voltage $v = 228{,}000$ V (4% above the rated 220kV of the transmission line). The control parameters are $k_1 = 200{,}000$ and $k_2 = 60{,}000$.

The initial demand for power is $\hat{p}_2 = (-500 + 50i) \times 10^6$ Watts. The initial conditions for the model are calculated numerically so that all time derivatives are zero. At $t = 0.1$ the demand changes abruptly to $\hat{p}_2 = (-525 + 50i) \times 10^6$ and this 5% change in real demand sets the system into motion. The load supplies negative power because the source of power points toward the zero junction. For the same reason, the generator supplies positive power.

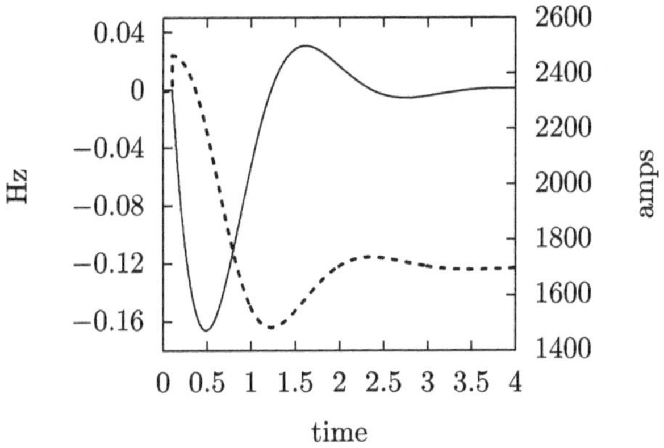

Figure 18. The change in frequency $2\pi\omega$ (solid line) and real current $\Re\{\hat{f}_1\}$ (dashed line) transferred through the line in response to a sudden change in real demand for power.

Figure 18 shows the simulation traces of $2\pi\omega$ and the real part of the current $\hat{f}_1$ flowing down the line. Following its initial jump, the real current fluctuates with the changing angle of the generator. Incidentally, this demonstrates that differences in phase angle are the mechanism for transferring real power in an alternating current system.

Exercises

Ex. 9.1 Replace the F element with an E element in the filter example. What frequencies does it pass?

Ex. 9.2 Construct electrical and translating mechanical devices that realize both filters. Hint: you will want an SE element to model ground with $e = 0$. The output signal will be the effort across the element connected to this ground.

Ex. 9.3 Design a *band pass* filter that removes the high and low frequency components of a signal but passes through a frequency in between these.

Ex. 9.4 Define the admittance matrix Y for a power system as follows. The entry y_{jk} is the admittance of the line from the jth to the kth zero junction. The diagonal element $y_{kk} = i2\pi\nu C_k + \sum_{j\in J} y_{jk}$. Define the vector $p = [\hat{p}_1 \dots \hat{p}_n]$ of power injected at each zero junction and similarly the vector $e = [\hat{e}_1 \dots \hat{e}_n]$ of efforts at each zero junction. Show that power and effort at each node are related by

$$\hat{p}_k = |\hat{e}_k| \sum_{j=1}^{n} |\hat{e}_j| e^{i(\theta_k - \theta_j)} y_{kj}$$

Ex. 9.5 Let $f(t)$ be a function with period T. Begin with the complex Fourier series

$$f(t) = \sum_{k=-\infty}^{\infty} c_k \exp(i2\pi kt/T)$$

where the c_k are complex numbers and derive the phasor.

10. Solutions to exercises

Ex. 1.1

$$\dot{e}_1 = f_2/E$$
$$\dot{f}_2 = -(e_1 + Zf_2)/F$$

Ex. 1.2

$$\dot{e}_1 = f_1/E$$
$$\dot{f}_2 = e_1/F$$
$$f_1 + e_1/Z + f_2 = 0$$

Ex. 1.3 f_2 is a known source of flow and

$$\dot{e}_1 = f_2/E$$
$$\dot{f}_7 = (e_6 - Z_3 f_7)/F$$
$$e_6 = Z_2(f_2 - f_7)$$
$$e_3 = Z_1 f_2$$

Ex. 1.4 e_1 is a known source of effort and

$$\dot{e}_7 = (f_8 + f_5)/E$$
$$\dot{f}_8 = (e_1 - e_7)/F$$
$$f_5 = (e_1 - e_7)/Z$$

Ex. 1.5 f_1 is a known source of flow and

$$\dot{e}_8 = (f_1 - f_7)/E$$
$$e_4 = (f_1 - f_7)Z$$
$$\dot{f}_7 = (e_4 + e_8)/F$$

Ex. 1.6

$$\dot{f}_2 = (e_1 - f_2 Z_1 - e_4)/F_1$$
$$\dot{e}_6 = (f_5 - f_9)/E$$
$$\dot{f}_9 = (e_6 - f_9 Z_2)/F$$
$$e_6 = G f_2$$
$$f_5 = f_6 + f_9$$
$$e_4 = G f_5$$
$$e_1 = e_2 + Z_1 f_2 + e_4$$

Ex. 1.7

$$\dot{e}_7 = f_7/E$$
$$e_7 = Z_3(f_5 - f_7)$$
$$e_2 - e_7 = Z_2 f_5$$
$$e_2 = Z_1(f_1 - f_5)$$

Ex. 1.8

$$\dot{f}_2 = (e_1 - Z_1 f_2 - T e_6)/F_1$$
$$\dot{e}_6 = (T f_2 - f_9)/E$$
$$\dot{f}_9 = (e_6 - f_9 Z_2)/F_2$$

Ex. 1.9 You should find that

$$e_1 - e_{12} = (Z_1 + Z_4) f_3 = Z_3 f_2 = (Z_2 + Z_5) f_4$$
$$e_1 - e_6 = f_4 Z_2$$
$$e_6 - e_{12} = f_4 Z_5$$
$$e_1 - e_5 = f_3 Z_1$$
$$e_5 - e_{12} = f_3 Z_4$$
$$f_1 + f_{12} = 0$$

Ex. 2.1 $f_1 = f_2$ and $f_4 = f_5$; $e_1 = e_2$ and $e_4 = e_5$. Therefore the zero junctions are redundant and can be removed.

Ex. 2.3

$$\dot{e}_5 = (f_3 - e_5/Z)/E$$
$$f_3 = I(\exp(\alpha(e_1 - e_5)) - 1)$$

The diode causes the flow f_3 to quickly approach zero as e_1 approaches e_5. When the flow is small, accumulated flow in the E element sustains e_5. When the diode flow is large, the E element accumulates this flow. If the parameters are selected carefully, then e_5 oscillates with small deviations about some desired constant.

Ex. 2.4 For a source of flow:

$$\dot{e}_3 = f_1/E$$
$$e_2 = Mq_2f_1$$
$$\dot{q}_2 = f_1$$

The effort e_2 will grow indefinitely. The physical device being modeled has some limit to the effort it can tolerate, and we can expect the device to fail when this limit is exceeded. For a source of effort:

$$\dot{e}_3 = f_1/E$$
$$f_1 = e_2/Mq_2$$
$$\dot{q}_2 = f_1$$
$$e_2 = e_1 - e_3$$

The flow will slow to a trickle as q_2 increases, and the system will approach an equilibrium state wherein $\dot{e}_3 \approx 0$.

Ex. 2.11 The graph has a single one junction attached to the two sources of effort and a single Z element. The Z element has parameter $1/Z = 1/(Z_1 + Z_4) + 1/Z_3 + 1/(Z_2 + Z_5)$

Ex. 3.6 The units of e are kg·m/s^2.

Ex. 4.1 The mass of fluid in the pipe is $\rho\ell A$. Therefore

$$\text{force} = \rho\ell A \cdot \text{acceleration} = \rho\ell \cdot \text{volume acceleration}$$

Pressure is force/area and it follows that $F = \rho\ell/A$. The units of $\rho\ell/A$ are kg/m^4.

Ex. 4.2 The graph is

$$\begin{array}{ccccc} & & \mathrm{F} & & \\ & & \uparrow & & \\ \mathrm{SE} & \xrightarrow{0} & 1 & \longrightarrow & \mathrm{Z} \end{array}$$

$$\dot{f} = (e_0 - Zf)/F$$

Ex. 4.3 With the pump, the graph is

$$\begin{array}{ccccc} & & \mathrm{F} & & \mathrm{E} \\ & & \uparrow & & \uparrow \\ \mathrm{SF} & \xrightarrow{1} & 1 & \xrightarrow{2} & 0 \\ & & \downarrow & & \downarrow \\ & & \mathrm{Z}_1 & & \mathrm{Z}_2 \end{array}$$

The simplified equations are

$$\dot{e}_2 = (f_1 - e_2/Z_2)/E$$
$$F\dot{f}_1 = e_1 - e_2 - f_1 Z_1$$

Notice that f_1 and $\dot{f}_1$ are defined by the source of flow. The only state variable is e_1. To remove the pump simply remove the SF element. Let f be the flow at the one junction and e the effort at the zero junction. The equations become

$$F\dot{f} = -e - fZ_1$$
$$\dot{e} = (f - e/Z_2)/E$$

This system has two state variables and its equilibrium state is $f = e = 0$; that is, when the water stored in the pipe and tank has drained out of the system.

Ex. 4.4 For configuration (1) there is a single flow $f = f_4 = f_7 = f_9$ for the single path from the cold water source to the shower head. Your equation should be

$$(Z_A + Z + Z_C)f_5 = e_5 + e_4$$

For configuration (2) the flow $f_9 = 0$. The equations are

$$\begin{aligned}
Z_A f_5 &= e_5 - e_3 \\
Z f_3 &= e_3 - e_1 \\
Z_B f_6 &= e_6 - e_2 \\
Z f_2 &= e_2 - e_1 \\
Z f_7 &= e_3 - e_9 \\
Z f_8 &= e_2 - e_9 \\
f_7 + f_8 &= 0 \\
f_7 &= f_5 - f_3 \\
f_8 &= f_6 - f_2
\end{aligned}$$

Ex. 4.5 One solution is to add zero junctions that attach to the ends of edge 1 and edge 4 where the sources of effort are in the original model. The sources of effort are then attached to these new zero junctions. By changing the direction of edges 1 and 4, the flows f_1 and f_4 take on the desired sign when fluid is leaving the network.

Ex. 4.6 The graph for case (b) is shown below. To get the graph for case (a) remove the F elements.

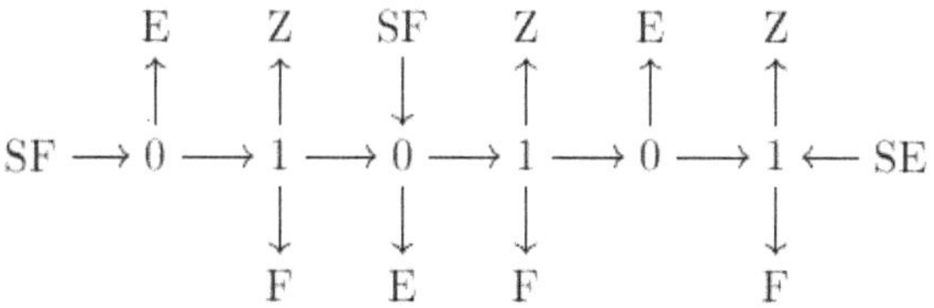

Ex. 5.1 Using τ for torque and F for force the relation is $\tau = RF$.

Ex. 5.2 We use efforts e_A, e_B and flows f_A, f_B for forces and velocities at the ends of each section. The relations are $Ae_A = Be_B$ and $Bf_A = Af_B$. If $B \gg$ A then a small force applied to the A segment produces a large force on the B segment. "Give me a lever long enough and a fulcrum on which to place it, and I shall move the world." – Archimedes of Syracuse

Ex. 5.3 The graph is

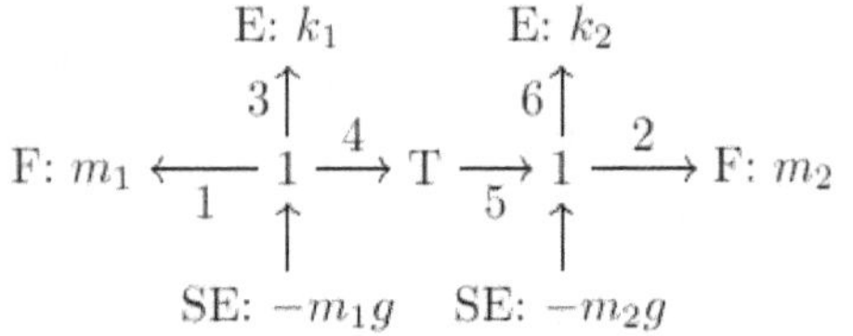

The positions of the masses are $y_1 = q_1$ and $y_2 = q_2$. The springs exert a downward force when they are compressed. The equations are

$$\begin{aligned}
\dot{f}_1 &= -(e_3 + m_1g + e_4)/m_1 \\
\dot{f}_2 &= (e_5 - e_6 - m_2g)/m_2 \\
\dot{e}_3 &= f_1/k_1 \\
\dot{e}_6 &= f_2/k_2 \\
Ae_4 &= Be_5
\end{aligned}$$

Ex. 5.4 We want the springs to exert an upward force when they are compressed. We accomplish this by retaining two of the extraneous zero junctions when simplifying the graph:

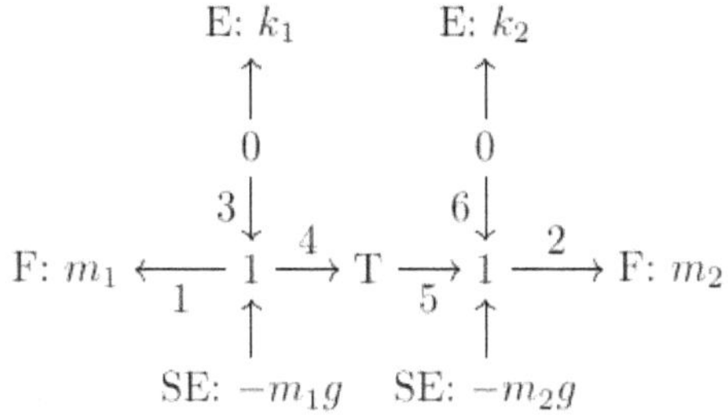

The equations are

$$
\begin{aligned}
\dot{f}_1 &= -(e_3 + m_1 g + e_4)/m_1 \\
\dot{f}_2 &= (e_5 - e_6 - m_2 g)/m_2 \\
\dot{e}_3 &= -f_1/k_1 \\
\dot{e}_6 &= -f_2/k_2 \\
Ae_4 &= Be_5
\end{aligned}
$$

Ex. 5.5 The graph is

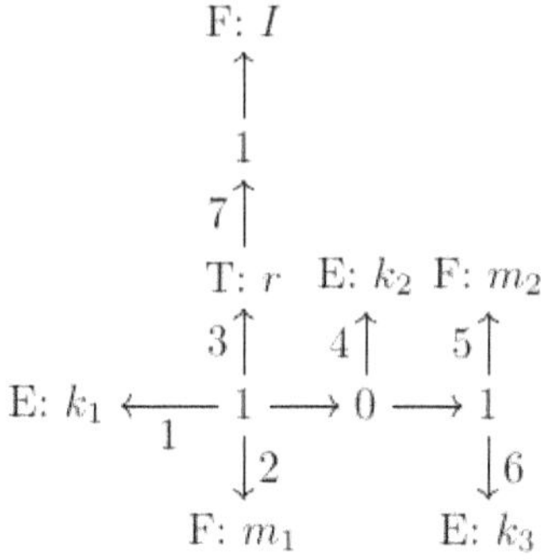

The equations are

$$\dot{e}_1 = f_2/k_1$$
$$\dot{f}_2 = -(e_1 + e_3 + e_4)/m_1$$
$$\dot{e}_4 = (f_2 - f_5)/k_2$$
$$\dot{e}_6 = f_5/k_3$$
$$\dot{f}_5 = (e_4 - e_6)/m_2$$
$$re_3 = e_7$$
$$rf_7 = e_3$$
$$I\dot{f}_7 = e_7$$

Ex. 5.6 Redundant edges are retained so that the velocities are simple to interpret as motion from left to right. When the spring is compressed, it opposes motion to the right; when extended, it pulls the mass to the right.

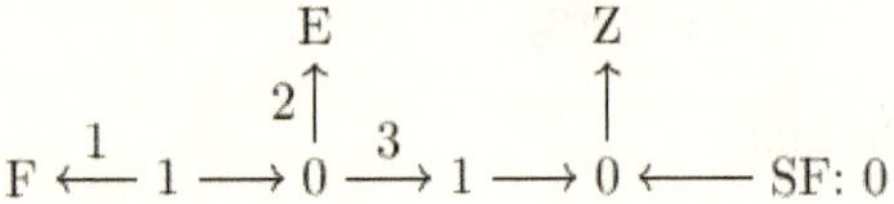

The equations are

$$\dot{f}_1 = -e_2/E$$
$$\dot{e}_2 = (f_1 - f_3)/F$$
$$f_3 = e_2/Z$$

Ex. 5.7 The graph becomes

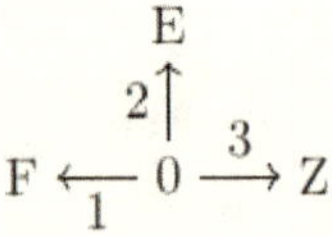

The equations are

$$\dot{f}_1 = e_2/F$$
$$\dot{e}_2 = -(f_1 + f_3)/E$$
$$f_3 = e_2/Z$$

If $f_1 > 0$ and $f_3 > 0$ then the spring is being compressed and the mass will (eventually) be pushed to the left; if they are both negative then the spring is being extended and the mass will (eventually) be pulled to the right. Therefore, if we interpret $f_1 > 0$ as motion from left to right, then $f_3 > 0$ must be motion from right to left.

Ex. 5.8 Attach an F element to the one junction adjacent to edge 3. The simplified graph is shown below; the velocity of F_2 has the same orientation as the point of connection that it replaces.

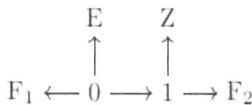

$$\begin{array}{ccccccc} & & \mathrm{E} & & \mathrm{Z} & & \\ & & \uparrow & & \uparrow & & \\ \mathrm{F}_1 & \leftarrow & 0 & \longrightarrow & 1 & \longrightarrow & \mathrm{F}_2 \end{array}$$

Ex. 6.2 The graph is

$$\begin{array}{ccccc} & & \mathrm{Z}\colon R & & \\ & & \uparrow & & \\ \mathrm{SF} & \longrightarrow & 0 & \longrightarrow & \mathrm{E}\colon C \\ & & \downarrow & & \\ & & \mathrm{F}\colon L & & \end{array}$$

Ex. 6.3 The graph is

$$\begin{array}{ccccccccc} & & \mathrm{F}\colon L_1 & & & & \mathrm{Z}\colon R_2 & & \\ & & \uparrow & & & & \uparrow & & \\ \mathrm{SE}\colon v_0 & \longrightarrow & 1 & \longrightarrow & 0 & \longleftarrow & 1 & \longleftarrow & \mathrm{SE}\colon v_1 \\ & & \downarrow & & \downarrow & & & & \\ & & \mathrm{Z}\colon R_1 & & \mathrm{F}\colon L_2 & & & & \end{array}$$

Ex. 6.4 The graph is

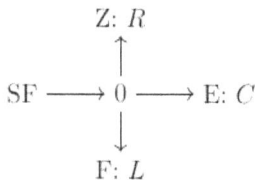

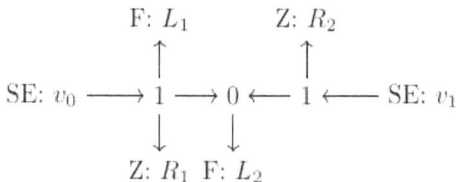

$$
\begin{array}{ccccccccc}
 & & \mathrm{F}\colon L_1 & & & & \mathrm{Z}\colon R_2 & & \\
 & & \uparrow & & & & \uparrow & & \\
\mathrm{SE}\colon v_0 & \longrightarrow & 1 & \longrightarrow & 0 & \longrightarrow & 1 & \longrightarrow & 0 \longleftarrow \mathrm{SE}\colon v_1 \\
 & & \downarrow & & \downarrow & & & & \\
 & & \mathrm{Z}\colon R_1 & & \mathrm{F}\colon L_2 & & & &
\end{array}
$$

Ex. 6.8 See the graph of the car body suspension in the chapter on mechanical systems. The source of effort has voltage $-m_1 g$ and the source of flow has current v. Currents are oriented just as velocities in the mechanical drawing.

Ex. 6.9 The graph is

$$
\begin{array}{ccccccccc}
 & & \mathrm{F}_1 & & & & & & \mathrm{Z} \\
 & & \uparrow & & & & & & \uparrow \\
\mathrm{SE} & \longrightarrow & 1 & \longrightarrow & 0 & \longrightarrow & \mathrm{T} & \longrightarrow & 0 \longrightarrow \mathrm{F}_2 \\
 & & & & \downarrow & & & & \\
 & & & & \mathrm{E} & & & &
\end{array}
$$

Ex. 7.1 The crank shaft and wheel act as a nonlinear transformer. Let $\theta = 0$ be along the $x -$ axis with increasing angle in the clockwise direction. The distance x from the center of the wheel to the leftmost connecting point satisfies the cosine law

$$l^2 = r^2 + x^2 - 2rx\cos\theta$$

Taking the time derivative we find

$$0 = 2x\dot{x} - 2r\dot{x}\cos\theta + 2rx\dot{\theta}\sin\theta$$

where $\dot{\theta} = \omega$ and the linear velocity $v = \dot{x}$. The equations for the transformer can be completed by simply requiring conservation of power

$$Fv = \tau\omega$$

Ex. 7.2 The graph is shown below. The flow f_0 is the angular speed of the wheel. The transformer T_1 is the crank shaft and T_2 the piston. The effort e_3 is the pressure in the piston chamber. The lower source of effort is the reservoir and the upper source of effort the pipe opening onto the atmosphere.

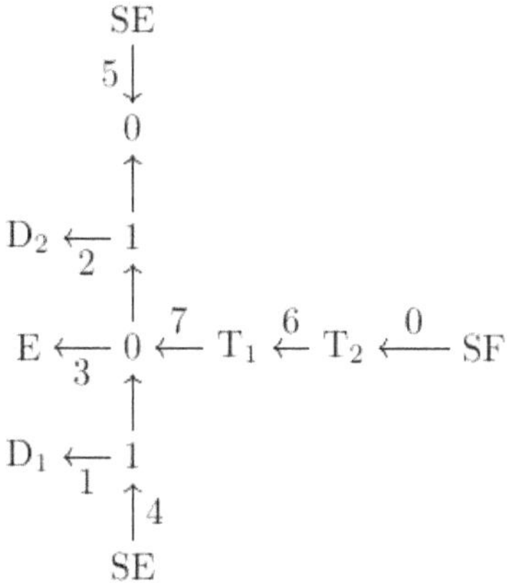

The equations for this graph are

$$
\begin{aligned}
&\dot{e}_3 = (f_1 + f_7 - f_2)/E\\
&0 = q_6 f_6 - r f_6 \cos q_0 + r q_6 f_0 \sin q_0\\
&f_6 e_6 = f_0 e_0\\
&f_7 = A f_6\\
&A e_3 = e_6\\
&f_1 = I_1(\exp(\alpha_1(e_4 - e_3) - 1)\\
&f_2 = I_2(\exp(\alpha_2(e_3 - e_5) - 1)\\
&\dot{q}_0 = f_0\\
&\dot{q}_6 = f_6
\end{aligned}
$$

Ex. 7.3 The piston changes direction in each rotation. The piston velocity $f_6 = 0$ at a change in direction. If $f_0 e_0$ is a positive constant, then this implies $f_6 e_6$ is also a positive constant, which is impossible if $f_6 = 0$.

Ex. 7.6 There are two sections of line. One is to the left of the wheel that is anchored to the ceiling. The other consists of the entire length of cable to the right of the wheel that is anchored to the ceiling. Suppose the pull force displaces the left section of line by a distance y. The right section of line must be displaced an equal distance. If we imagine the right section to consist of two equal parts separated by the movable block, then each must be displaced by a length $y/2$. Hence, the movable block is displaced by a distance $y/2$. Letting v_1 be the velocity of the pull line and v_2 the velocity of the movable block, we have

$$v_1 = 2v_2$$

It follows that the pull force F_1 and lift force F_2 are related by

$$2F_1 = F_2$$

So that a downward pull with force F_1 on the line results in an upward force $2F_1$ acting on the load.

Ex. 7.7 The graph is

$$\begin{array}{ccccccc}
& \text{F: } L & & & & \text{F: } m \\
& \uparrow & & & & \uparrow \\
\text{SE} \xrightarrow{1} & 1 & \longrightarrow \text{G} \longrightarrow 1 \longrightarrow \text{T} & \xrightarrow{2} & 1 \\
& & & & & \uparrow \\
& & & & & \text{SE: } -mg
\end{array}$$

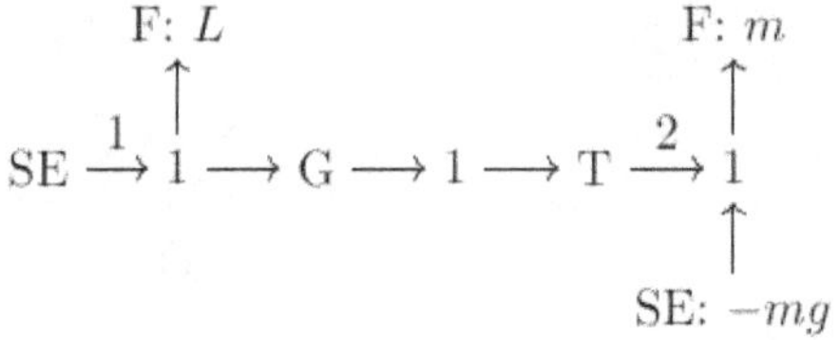

The equations for this graph are

$$\dot{f}_1 = (e_1 - (G/T)f_2)/L$$
$$\dot{f}_2 = ((G/T)f_1 - mg)/m$$

Ex. 8.1 The hot and cold sides are sources of effort. Heat flows from the hot side to the cold side. The graph is

$$\mathrm{SE} \xrightarrow{1} \mathrm{R}_1 \xrightarrow{2} 1 \xrightarrow{3} \mathrm{R}_2 \xrightarrow{4} 0 \longleftarrow \mathrm{SE}$$

The equations are

$$\begin{aligned}
&e_1 - e_2 = R_1 e_1 f_1 = R e_2 f_2 \\
&e_2 = e_3 \\
&f_2 = f_3 \\
&e_3 - e_4 = R_2 e_3 f_3 = R e_4 f_4
\end{aligned}$$

From these it follows that

$$\begin{aligned}
&e_1 - e_4 = (R_1 + R_2) e_1 f_1 \\
&e_1 f_1 = e_4 f_4
\end{aligned}$$

Ex. 8.2 A source of flow models the turning of the drill by the fire maker. A source of effort models the ambient temperature. The other elements should be self-explanatory.

$$\begin{array}{c}
\mathrm{SF} \\
\downarrow 1 \\
\mathrm{ZS}\!: KF \\
\downarrow 2 \\
\mathrm{C} \xleftarrow[3]{} 0 \xrightarrow[4]{} \mathrm{R} \xrightarrow[5]{} 0 \longleftarrow \mathrm{SE}
\end{array}$$

The equations for this graph are

$$\begin{aligned}
&C\dot{e}_3 = (f_2 - f_4) e_3 \\
&e_4 f_4 = e_5 f_5 \\
&e_4 - e_5 = R e_4 f_4 \\
&e_1 f_1 = e_2 f_2 \\
&f_2 = f_3 + f_4 \\
&e_2 = e_3 = e_4 \\
&KF f_1 = e_1
\end{aligned}$$

Ex. 8.3 We sketch a solution using pseudo graph elements (see Exercise 8.6). The total heat power entering the forming ember is $KFf_1^2 - (e_3 - e_5)/R$. Therefore

$$C\dot{e}_3 = -e_3/R + e_5/R + KFf_1^2$$

All terms except e_3 are constant and so we can solve the differential equation (analytically or numerically) and find the time t at which $e_3(t) = T$. The total energy expended by the fire maker to turn the drill (and, believe me, this is where you do the work!) is $KFf_1^2 t$. Softwoods have a higher thermal resistance and lower thermal capacitance compared to hardwoods for a fixed volume.[15] Therefore, softwoods waste less of the heat we supply by friction, burn more easily, and we should expect softwoods to be preferred when making fires (which, indeed, they are).

Ex. 8.4 The graph for a two element model is shown below.

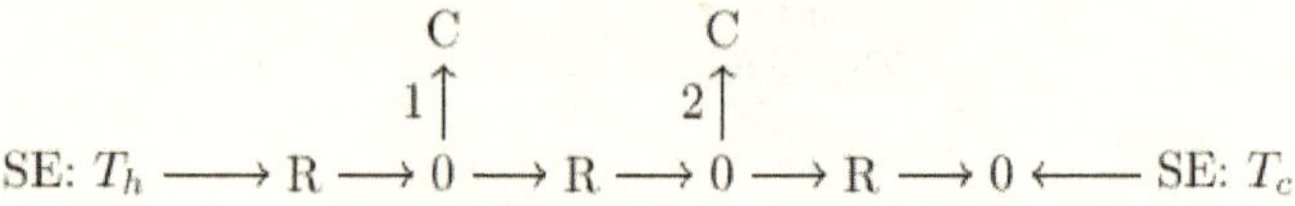

Consider just the temperature T_1. Its evolution can be described by

$$\dot{T}_1 = \frac{1}{RCh^2}(T_h - 2T_1 + T_2)$$

which, if we consider time derivatives at every C element, is a second order, finite difference approximation to

[15] W.P. Goss and R.G. Miller, *Thermal Properties of Wood and Wood Products*, in Thermal Performance of the Exterior Envelopes of Whole Buildings XIII International Conference, December 4-8, 2016.

$$\frac{\partial}{\partial t}T = \frac{1}{RC}\frac{\partial^2}{\partial x^2}T$$

Ex. 8.5 The ambient temperature is a source of effort. The directions of the R_3 edges are entirely arbitrary.

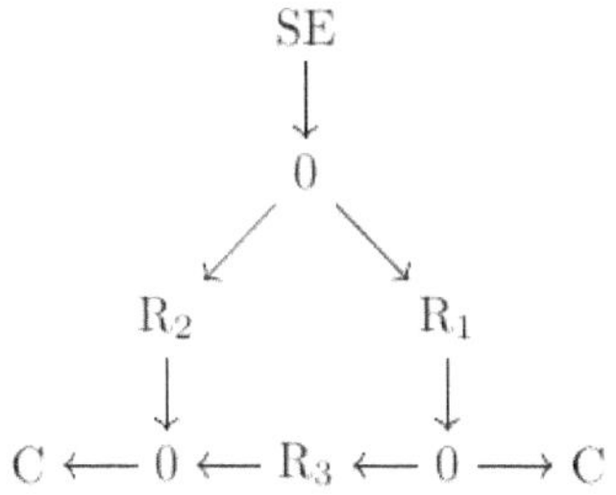

Ex. 8.6 The one junction has equal power and the temperatures sum to zero. The zero junction has equal temperatures and the powers sum to zero. The element relations are

$$C\dot{T} = P \text{ for the C element}$$
$$T_1 - T_2 = RP \text{ for the R element}$$
$$P_1 = P_2 + e_3 f_3 \text{ for the W element}$$
$$e_3 f_3 = P_1 \eta$$

Ex. 8.7 A source of effort is used for the outside air temperature T_a. The building insulation from the outside air is R_1 and heat transfer between rooms is modeled with element R_2. The rooms are C_1 and C_2. The W element is driven by a source of power (from the electrical, mechanical, or some other domain).

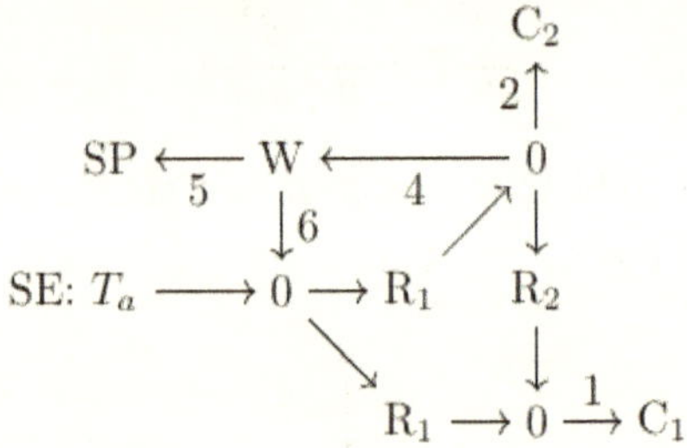

The equations for this model are

$$
\begin{aligned}
\dot{e}_1 &= P_1/C_1 \\
\dot{e}_2 &= P_2/C_2 \\
P_1 &= (T_a - e_1)/R_1 + (e_1 - e_2)/R_2 \\
P_2 &= (T_a - e_2)/R_1 - (e_1 - e_2)/R_2 - P_4 \\
P_4 &= P_6 + e_5 f_5 \\
e_5 f_5 &= \eta P_4
\end{aligned}
$$

The driving power $e_5 p_5$ and ambient temperature T_a are given. The unknowns are e_1, e_2, P_1, P_2, P_4, and P_6. Hence, we have six equations in six unknowns.

Ex. 9.1 The input to output function of the filter is

$$\hat{e}_1 = \left(Z + i2\pi\nu\frac{E}{Z}\right)\hat{e}_2$$

This is a low pass filter. Notice that the complex resistance grows with ν.

Ex. 9.2 The electric circuits and the low pass mechanical device are shown below. The high pass device has a mass in place of the spring.

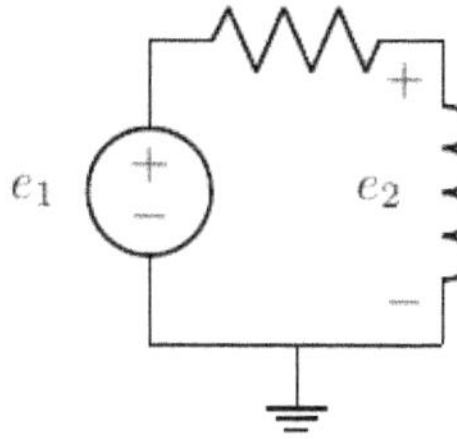

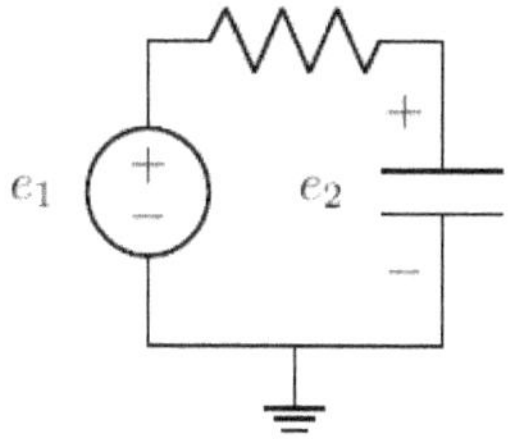

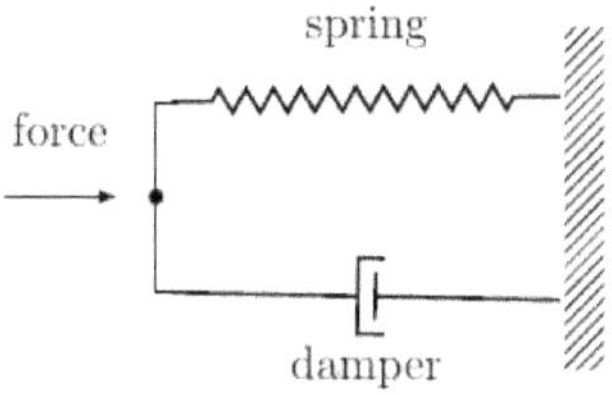

Ex. 9.3 This will consist of a high pass filter connected to a low pass filter. One possible (but not the only) solution is shown below. The input is e_1 and the output is e_2. The Z_1 and F elements act as a high pass filter. The Z_2 and E elements act as a low pass filter.

$$\begin{array}{ccccc} Z_1 & & F & & Z_2 \\ \uparrow & & \uparrow & & \uparrow \\ SE \xrightarrow{1} 1 & \longrightarrow & 0 & \longrightarrow & 1 \xrightarrow{2} E \end{array}$$

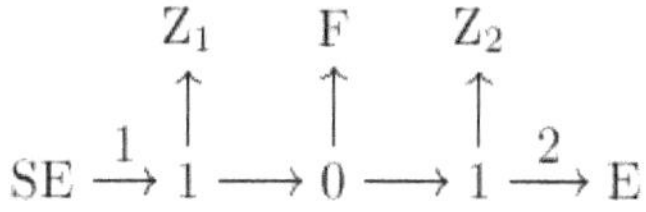

www.ingramcontent.com/pod-product-compliance
Lightning Source LLC
Chambersburg PA
CBHW021559210326
41599CB00010B/519
* 9 7 9 8 3 3 7 0 8 9 5 4 6 *